Holt Pre-Algebra

Family Involvement Activities

HOLT, RINEHART AND WINSTON

A Harcourt Education Company

Orlando • **Austin** • New York • San Diego • Toronto • London

Printed in the United States of America

ISBN 0-03-069697-6

3 4 5 6   082   06 05 04

CONTENTS

Holt Pre-Algebra

CONTENTS, *CONTINUED*

Holt Pre-Algebra

What We Are Learning

Equations and Inequalities

Vocabulary
These are the math words we are learning:

Addition Property of Equality Add the same number to both sides of an equation, and the new equation will have the same solution.

Algebraic Expression An expression with one or more variables.

Algebraic Inequality An inequality that contains a variable.

Coefficient The number multiplied by the variable.

Constant A value that does not change.

Division Property of Equality Divide both sides of the equation by the same number and the new equation will have the same solution.

Equation A mathematical sentence that contains an equal sign.

Equivalent Expression Expressions that have the same value for all the values of the variables.

Evaluate To find the value of a numerical or algebraic expression.

Inequality A statement comparing expressions that are not equal.

Dear Family,

Your child will be learning about equations and inequalities. Initially, your child will be evaluating algebraic expressions that contain up to two variables. When given the values of the variables, your child will simply "substitute" those values for the intended variable and then simplify the expression. This skill will help your child check their answers when solving equations.

Another skill your child will be reviewing is writing algebraic expressions in place of word phrases. Listed below are examples of the four basic operations and some key phrases that may help your child write these algebraic expressions.

Operation	Key Word Phrases	Expression
Addition	a number plus 6 add 6 to a number the sum of 6 and a number 6 more than a number	$6 + m$
Subtraction	a number minus 3 3 less than a number a number decreased by 3 the difference of a number and 3	$m - 3$
Multiplication	8 times a number the product of a number and 8	$8m$
Division	a number divided by 2 2 divided into a number the quotient of a number and 2	$m \div 2$ or $\frac{m}{2}$

Your child needs to be very familiar with the four basic operations and how each operation works. This is a basic building block for solving equations, another concept your child will be learning in this section.

To effectively solve equations, your child will learn the four equality properties: addition, subtraction, multiplication, and division. These properties are also known as inverse operations. Your child will use these properties to isolate the variable in order to solve an equation.

An important element to equation solving is for students to always check the solution. As you help your child with his or her homework, make sure he or she does not forget this very important step.

Holt Pre-Algebra

Inverse Operation
Operations that "undo" each other.

Isolate the Variable
Getting the variable alone on one side of the equal sign.

Like Term Terms that have the same variable raised to the same power.

Multiplication Property of Equality Multiply both sides of an equation by the same number and the statement will still be true.

Simplify Perform all the operations possible, including combining like terms.

Solution The value of the variable that makes the equation true.

Solution of an Inequality A number that makes an inequality true.

Solution Set The set of all solutions that make the statement true.

Solve Find the value of the variable that makes the equation true.

Subtraction Property of Equality Subtract the same number from both sides of the equation and the statement will still be true.

Term Elements of an expression that are separated by plus or minus signs.

Variable A letter that represents a value that can change or vary.

This is how your child will use the properties of equality to solve a simple two-step equation.

Solve $4x + 3 = 15$.

$$4x + 3 = 15$$
$$\underline{ - 3 \quad -3}$$
$$4x = 12$$

Subtract 3 from both sides using the Subtraction Property of Equality to isolate the x-term.

$$\frac{4x}{4} = \frac{12}{4}$$
$$x = 3$$

Divide both sides by 4 using the Division Property of Equality in order to completely isolate the variable, x.

Check:

$$4x + 3 \stackrel{?}{=} 15$$
$$4(3) + 3 \stackrel{?}{=} 15$$
$$12 + 3 \stackrel{?}{=} 15$$
$$15 = 15 \ ✓$$

Substitute 3 for x.

After equation solving, your child will transition to solving and graphing inequalities. The same steps will be followed to isolate a variable in an inequality as with an equation. The difference will be in the solution set. Your child will graph the solution to the inequality on a number line.

Finally, your child will learn how to combine like terms in order to simplify an expression. Combining like terms is an important concept for your child to grasp. Your child will learn that just because he or she changes the way an expression looks, the value of the expression stays the same. For example:

Combine like terms.
$6x + 8y + 9x$

$6x + 8y + 9x$	Identify like terms.
$6x + 9x + 8y$	Rearrange the expression.
$15x + 8y$	Add the coefficients of like terms.

This review of algebraic concepts will give your child the confidence and skills needed to be successful in the chapters to follow.

Sincerely,

Holt Pre-Algebra

CHAPTER	**Family Letter**
1	*Equations and Inequalities*

Evaluate each expression for the given value(s) of the variable.

1. $8(x + 7)$ for $x = 17$

2. $6m - 7n$ for $m = 12$ and $n = 7$

Write an algebraic expression for each word phrase.

3. a number decreased by 5

4. the product of a number and 9, less 4

Determine which value of the variable is a solution of the equation.

5. $c + 5 = 61$; $c = 46$, 56, or 66

6. $d - 37 = 12$; $d = 39$, 49, or 59

Solve.

7. $19 = w - 4$

8. $3x = 42$

9. $3n + 2 = 11$

10. $\frac{p}{7} = 9$

11. $8 = a - 7$

12. $16s = 48$

Use $<$ or $>$ to complete each inequality.

13. $3 + 5 \boxed{} 10$

14. $(3)5 \boxed{} 16$

15. $6.1 \boxed{} 1.5(4)$

Combine like terms.

16. $3x + 6v + 5x - 3v + 8$

17. $4s - 3s + 2 + 5j + 9s + 7j$

Simplify.

18. $8(5 + s) - 2s$

19. $2(4y - 2) + 7$

Solve.

20. $5w + 3w = 16$

21. $12x - 7x = 25$

CHAPTER 1

Family Fun
Hidden Math Terms

Directions
Find the hidden math terms from this section. The terms may be written backwards, forwards, or on a diagonal. For fun, make copies of the puzzle and give it to a friend or family member. Time how long it takes each person to find all of the terms.

```
L I X I Z B U A T A Q N E A E T H A N G E E
K G X N Q U Z J S Q U H C Y Y X C L I S E L
I N V E R S E O P E R A T I O N S G V S B B
S U O Q P G L W H T E C R F I J B E R H U A
G F S U P V T T G T E T W Y B O K B D O A I
H V O A E C C N A B Q X V Z U X R R W Z F R
L V V L T Y O U A N M W L L J D B A N T J A
E Q U I V A L E N T E X P R E S S I O N S V
V J X T Y A D L F S S Z A H B A O C I V F E
C Q R Y V E I X U F E N W G R U P E T A I H
Q I S E Y P F B P I I D O L Q H E X A R Y T
O L F O N D S G N P Z C Q C W H W P U I C E
K M I V L T W S I M P L I F Y D M R Q A M T
D A Z R I U V Q S M R E T E K I L E E B N A
C G F T K Y T S G F C W C Z N F M S U L P L
M I U H H E L I U K V N N R I T S S H E H O
P T N B S G Y A O L T B J W I O C I A C Y S
E H U C L K V A R N X V V I F O V O Y V R I
Y I Q L F Y S V P N B U X C V Z E N H S J B
Y T I L A U Q E N I C I A R B E G L A N N P
```

ALGEBRAIC EXPRESSION	**INVERSE OPERATIONS**
ALGEBRAIC INEQUALITY	**ISOLATE THE VARIABLE**
COEFFICIENT	**LIKE TERMS**
CONSTANT	**SIMPLIFY**
EQUATION	**SOLUTION**
EQUIVALENT EXPRESSIONS	**SOLVE**
EVALUATE	**SUBSTITUTE**
INEQUALITY	**VARIABLE**

Holt Pre-Algebra

What We Are Learning

Graphing

Vocabulary
These are the math words we are learning:

Coordinate Plane A plane formed by two number lines that intersect at right angles.

Graph of an Equation The set of all ordered pairs that are solutions of the equation.

Ordered Pair A pair of numbers that can be used to locate a point on a coordinate plane.

Origin The point (0, 0), which is at the intersection of the *x*- and *y*-axis.

x-axis The horizontal number line on the coordinate plane.

x-coordinate The first number in an ordered pair.

y-axis The vertical number line on the coordinate plane.

y-coordinate The second number in an ordered pair.

Dear Family,

Now that your child is familiar with solving equations, your child will learn how to write solutions of equations that have two variables. When an equation contains two variables, the solution is written as an **ordered pair,** (*x*, *y*).

When given an ordered pair as a solution for an equation, this is how your child will decide if it is in fact a solution of the given equation.

Determine if the ordered pair (2, 7) is a solution of $y = 3x + 3$.

$y = 3x + 3$	Substitute the values for *x* and *y*.
$7 = 3(2) + 3$	Multiply.
$7 \neq 9$	Add.

The ordered pair (2, 7) is *not* a solution to the equation.

Your child will also learn to create a table of ordered pairs for a given equation. Students will be given an equation and then a list of values for one of the variables. By substituting the value for the indicated variable, your child will be able to determine the value for the second variable.

Make a table of solutions for $x = 1, 2, 3, 4$.
$y = 8x - 4$

x	1	2	3	4
$8x - 4$	$8(1) - 4$	$8(2) - 4$	$8(3) - 4$	$8(4) - 4$
y	4	12	20	28
(*x, y*)	(1, 4)	(2, 12)	(3, 20)	(4, 28)

Your child will use ordered pairs to plot points on a coordinate graph. An ordered pair is a location on a coordinate grid. The **x-coordinate** is always the first value in the ordered pair. It tells you how many units to move from the origin left or right. The **y-coordinate** is the second value of the ordered pair. It tells you how many units to move from the origin up or down.

Holt Pre-Algebra

Once your child is proficient at plotting points on a coordinate grid, the instruction will transition to graphing equations of lines.

Here is how your child will graph the equation $y = 8x - 4$.

Plot any two ordered pairs from the table of solutions shown on the previous page, i.e. (1, 4) and (2, 12).

To plot (1, 4) begin at the origin and move right 1 unit and up 4 units.
To plot (2, 12) begin at the origin and move right 2 units and up 12 units.

Then connect the two points with a straight line.

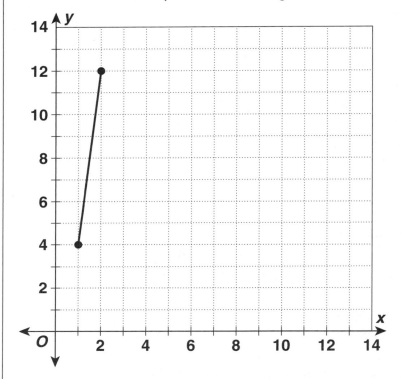

Have your child verbalize their interpretation of the graph display.

Encourage your child to complete their daily homework assignments. Mastering these concepts is essential for success in this chapter.

Sincerely,

Holt Pre-Algebra

Name _____ Date _____ Class _____

Determine if the ordered pair is a solution of $y = 12x + 2$**.**

1. (1, 14) **2.** (0, 2) **3.** (3, 36) **4.** (2, 24)

_____ _____ _____ _____

Use the given values to complete the table of solutions.

5.

$y = 7x + 2$	
x	y
0	
1	
2	
3	
4	

6.

$y = 6x - 4$	
x	y
1	
3	
5	
7	
9	

Use the coordinate plane for Exercises 7 and 8.

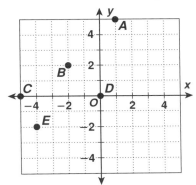

Give the coordinate of each point.

7. A _____ B _____ C _____

   D _____ E _____

Graph each ordered pair on the coordinate plane. Label points *F*, *G*, and *H*.

8. $F = (4, 3)$; $G = (-1, 0)$; $H = (4, -3)$

Complete the table of ordered pairs. Graph the equation on the coordinate plane.

9.

$y = 5x + 3$			
x	5x + 3	y	(x, y)
0			
1			
2			

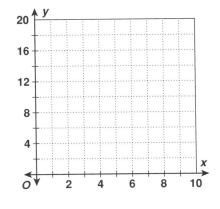

line should be drawn from (0, 3) to (2, 13).
unit left; Point *H* is located 4 units right, 3 units down. 9. 3, 3, (0, 3); 8, 8, (1, 8); 13, 13, (2, 13); A straight
$C = (-5, 0)$, $D = (0, 0)$ $E = (-4, -2)$ 8. Point *F* is located 4 units right, 3 units up; Point *G* is located 1
(2) $B = (-2, 2)$
Answers: 1. yes 2. yes 3. no 4. no 5. 2, 9, 16, 23, 30 6. 2, 14, 26, 38, 50 7. $A = (1, 5)$

Holt Pre-Algebra

Name _____ Date _____ Class _____

Why did the chicken lose its head?
Because it did not _____

Directions

Plot the points for the ordered pairs in the order given. Connect the points
with straight lines to reveal the answer to the riddle.

(−5, 10)
(−2, 9)
(0, 8)
(2, 7)
(4, 5)
(5, 3)
(5, 0)
(3, −5)
(1, −7)
(−2, −9)
(−2, −12)
(−2, −15)
(−3, −13)
(−5, −12)
(−7, −11)
(−8, −9)
(−6, −7)
(−4, −6)
(−5, −3)
(−6, 0)
(−5, 3)
(−4, 5)
(−3, 7)
(−3, 8)
(−5, 10)

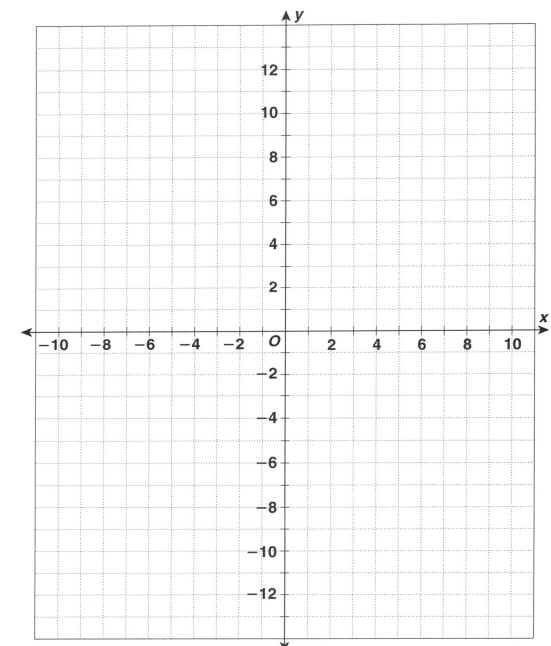

Answer: Duck

Holt Pre-Algebra

Family Letter

What We Are Learning

Integers

Vocabulary
These are the math words we are learning:

Absolute Value The distance a number is from zero on a number line.

Integers The set of whole numbers and their opposites.

Opposites Two numbers that are an equal distance from zero on a number line.

Dear Family,

Your child will be learning about a special set of numbers called integers. The set of integers includes the set of whole numbers and their opposites.

Number lines will be used to introduce the concept of "opposite" numbers. The opposite of a number is the same distance from 0, as the given number on a number line. In mathematics, this is called absolute value, designated by the symbol | |.

Absolute value is always positive; it denotes distance from zero, and distance cannot be negative.

$$|7| = 7 \qquad |-7| = 7$$

Your child will learn how to add, subtract, multiply, and divide integers by following a few important, yet simple, guidelines.

Guidelines for Adding Integers	
If the signs are the **same,** find the **sum** of the absolute values of the integers and give that sum the same sign of the integers in the problem.	If the signs are **different**, find the **difference** of the absolute values. Use the sign of the integer with the larger absolute value.
Add. $\quad -3 + -6$ *Think:* Find the sum of 3 and 6. The signs are the same so give the sum the sign of the integers.	**Add.** $\quad -2 + 5$ *Think:* Find the difference between 5 and 2. Because $5 > 2$, use the sign of 5.
$-3 + -6 = -9$	$-2 + 5 = 3$

The strategies used for adding integers will also be applied to subtracting integers. To subtract integers, your child will change the subtraction sign to an addition sign and then *add the opposite* of what is shown. This is how your child will subtract integers.

Subtract.

A. 6 − 3
$\quad 6 + (-3) = 3$ Add the opposite of 3. $6 > 3$, use the sign of 6.

B. −5 − 8
$\quad -5 + (-8) = -13$ Add the opposite of 8. Same sign, use the sign of the integers.

Holt Pre-Algebra

The process of multiplying and dividing integers is very similar to multiplying and dividing whole numbers, except that the product/quotient has a sign.

Students will learn these simple guidelines to determine the sign of their answer.

- If the integers have the same sign, the product or the quotient will **ALWAYS** be positive.

  $-4(-6)$ The signs are the same, so the answer is
  24   positive.

- If the integers have different signs, the product or quotient will **ALWAYS** be negative.

  $-4(6)$ The signs are different, so the answer is
  -24 negative.

Your child will use their knowledge of integer operations to solve basic integer equations. The process of solving equations with integers is the same as solving equations with whole numbers, to isolate the variable. The same is true when using integers to solve inequalities. However, one important step must be remembered:

When isolating the variable, reverse the inequality sign if you multiply or divide by a negative number.

Solve.

A. $\dfrac{a}{-3} > 6$

$(-3)\dfrac{a}{-3} < 6(-3)$ Multiply each side by -3, $>$ changes to $<$.

     $a < -18$

B. $-3a > 18$

   $\dfrac{-3a}{-3} < \dfrac{18}{-3}$ Divide each side by -3, $>$ changes to $<$.

     $a < -6$

This section will cover many fundamental concepts concerning integers and properties of integers. Practice mental math games with your child to help sharpen your child's skills with integers.

Sincerely,

Holt Pre-Algebra

Name _____ Date _____ Class _____

Family Letter

Integers

Add.

1. $-3 + 8$ **2.** $-4 + (-5)$ **3.** $6 + (-9)$ **4.** $12 + (-7)$

_____   _____   _____   _____

Evaluate each expression for the given value of the variable.

5. $t + 8$ for $t = -12$ **6.** $b + (-5)$ for $b = 3$ **7.** $x + 11$ for $x = -19$

_____   _____   _____

Subtract.

8. $-9 - 7$ **9.** $15 - (-3)$ **10.** $-19 - (-4)$ **11.** $32 - (-17)$

_____   _____   _____   _____

12. The temperature was 3° below zero at 6 P.M. Six hours later the temperature fell another 18°. What was the temperature at midnight?

_____

Multiply or divide.

13. $-3(6)$ **14.** $-4(-5)$ **15.** $\dfrac{60}{-12}$ **16.** $16(-2)$

_____   _____   _____   _____

17. $\dfrac{-75}{-3}$ **18.** $-9(-9)$ **19.** $\dfrac{-18}{3}$ **20.** $-2(-10)$

_____   _____   _____   _____

Simplify.

21. $12(8 - 12)$ **22.** $-6(-3 + 7)$ **23.** $11 + 3(4 - 9)$ **24.** $7 - 8(4 + 6)$

_____   _____   _____   _____

Solve.

25. $-6 + x = 15$ **26.** $-3a > -33$ **27.** $\dfrac{a}{15} = -2$ **28.** $f - 8 = -23$

_____   _____   _____   _____

Answers: 1. 5 **2.** −9 **3.** −3 **4.** 5 **5.** −4 **6.** −2 **7.** −8 **8.** −16 **9.** 18 **10.** −15 **11.** 49 **12.** −21° **13.** −18 **14.** 20 **15.** −5 **16.** −32 **17.** 25 **18.** 81 **19.** −6 **20.** 21 **21.** −48 **22.** −24 **23.** −4 **24.** −73 **25.** x = 21 **26.** a < 11 **27.** a = −30 **28.** f = −15

Holt Pre-Algebra

CHAPTER
2

Family Fun
Integer Cards

Materials
Deck of regular playing cards
Score pad and pencil

Directions

- Shuffle the cards. Each card has a value.
  - **All** of the black cards are positive integers.
  - **All** of the red cards are negative integers.

  The face cards have special qualities:
  - Jacks are multiplied by the value of 2.
  - Queens are multiplied by the value of 5.
  - Kings are multiplied by the value of 10.

- Each person starts out with 100 points.

- Each round consists of each player being dealt 2 cards, finding the sum of those integers, and then adding this integer to their total, beginning with 100. If a face card is dealt, the player must find the product of the two cards and then add the new integer to the player's total score.

- Record the total after each round under your name on the score pad.

- The goal is to be the player with the lowest positive score after 10 rounds.

Round				
1				
2				
3				
4				
5				
6				
7				
8				
9				
10				

Holt Pre-Algebra

Family Letter

Section B

What We Are Learning

Exponents

Vocabulary
These are the math words we are learning:

Base A number that is used as a factor in a power.

Exponent A number that represents how many times the base will be used as a factor in a power.

Power A number produced by raising a base to an exponent.

Scientific Notation A shorthand way of writing very large or very small numbers using powers of 10.

Dear Family,

Your child will be learning about exponents and the properties associated with them. An exponent is a part of a power. The exponent is a number that represents how many times the base is to be multiplied.

Using exponents will allow your child to write repeated multiplication in a more efficient way. This is how your child will write and evaluate exponents.

Write using exponents.
$2 \cdot 2 \cdot 2 \cdot 2$

$2 \cdot 2 \cdot 2 \cdot 2 = 2^4$ Identify how many times 2 is a factor.

Evaluate 4^3.

$4^3 = 4 \cdot 4 \cdot 4$ Find the product of three 4's.

$\quad = 64$

When evaluating expressions with exponents, remind your child to follow the order of operations.

As your child becomes familiar with writing and expressing exponents, they will learn how to apply exponent properties. There are important relationships that exist between exponents and the operations of multiplication and division. Your child will learn how to multiply and divide powers with the same base by using some of these basic properties of exponents.

Multiplying
When multiplying powers with the *same* base, keep the base constant and add the exponents.

Multiply. Write the product as one power.
$5^6 \cdot 5^2$

5^{6+2} Add the exponents.

5^8

Dividing
When dividing powers with the *same* base, keep the base constant and subtract the exponents.

Divide. Write the quotient as one power.
$\dfrac{12^{14}}{12^9}$

12^{14-9} Subtract the exponents.

12^5

Holt Pre-Algebra

When multiplying or dividing powers with the same base it is important to remember **not** to multiply or divide the bases. This property only applies when bases are the same. Powers with unlike bases cannot be combined using the two properties shown in the examples.

A special case occurs when the bases are the same and the difference in the exponents is zero. The zero power of any number, except zero, will always equal 1. For example, $5^0 = 1$.

Since it is possible to have a sum or difference that is a negative number, your child will learn how to evaluate expressions with negative exponents. A number raised to a negative exponent equals 1 divided by that number raised to the opposite of the exponent. This is how your child will learn to evaluate expressions with negative exponents.

Evaluate.

$\dfrac{5^4}{5^6}$

5^{-2} The bases are the same so subtract the exponents.

$\dfrac{1}{5^2}$ Write the reciprocal and change the sign of the exponent.

$\dfrac{1}{25}$ Simplify.

Have your child explain the purpose of exponents and the many ways they are used in mathematics. Allowing your child to verbalize this information is a way to help your child understand this material.

Sincerely,

Holt Pre-Algebra

Name _____ Date _____ Class _____

Family Letter
Exponents

Write using exponents.

1. $(-4) \cdot (-4) \cdot (-4)$ 2. $3 \cdot 3 \cdot 3 \cdot 3 \cdot 3 \cdot 3$ 3. $9 \cdot 9$ 4. $t \cdot t \cdot t$

_____   _____   _____   _____

Evaluate.

5. 4^3 6. $(-3)^4$ 7. 6^4 8. $(-1)^9$

_____   _____   _____   _____

Simplify.

9. $(2 + 4^3)$ 10. $(27 - 3^2)$ 11. $25 + (8 \cdot 4^3)$ 12. $(12 - 7^2)$

_____   _____   _____   _____

Multiply. Write the product as one power.

13. $6^4 \cdot 6^3$ 14. $5^3 \cdot 5^6$ 15. $8^2 \cdot 8^7$ 16. $k^8 \cdot k^2$

_____   _____   _____   _____

Divide. Write the quotient as one power.

17. $\dfrac{5^6}{5^4}$ 18. $\dfrac{r^7}{r^6}$ 19. $\dfrac{8^{12}}{8^6}$ 20. $\dfrac{p^8}{p^4}$

_____   _____   _____   _____

Evaluate the powers of 10.

21. 10^{-3} 22. 10^{-6} 23. 10^{-8} 24. 10^{-1}

_____   _____   _____   _____

Evaluate.

25. 3^{-3} 26. $\dfrac{6^2}{6^5}$ 27. $\dfrac{4^5}{4^3}$ 28. $7^4 \cdot 7^{-5}$

_____   _____   _____   _____

Write each number in standard or scientific notation.

29. $9{,}630{,}000$ 30. 2.7×10^{-5} 31. 0.0015

_____   _____   _____

Answers: 1. $(-4)^3$ 2. 3^6 3. 9^2 4. t^3 5. 64 6. 81 7. 1296 8. -1 9. 66 10. 18 11. 537 12. -37 13. 6^7 14. 5^9 15. 8^9 16. k^{10} 17. 5^2 18. r^1 19. 8^6 20. p^4 21. 0.001 22. 0.000001 23. 0.00000001 24. 0.1 25. $\frac{1}{27}$ 26. $\frac{1}{216}$ 27. 16 28. $\frac{1}{7}$ 29. 9.63×10^6 30. 0.000027 31. 1.5×10^{-3}

Holt Pre-Algebra

CHAPTER 2 **Family Fun**

Match It Up!

Directions

Using the numbers in the right-hand column, make the number sentences true in the left-hand column. You may use some numbers more than one time and some numbers not at all. See how well you can match up the correct answers to the number sentences.

Number Sentences	x equals _____	Possible Answers for x
1. $3^x = \frac{1}{9}$		0
2. $7.2 \times 10^x = 0.072$		10
3. $x^{12} = 1$		−5
4. $8^x \times 8^3 = 8^{-2}$		2
5. $x^4 = 10{,}000$		7
6. $\frac{7^8}{7^x} = 49$		3
7. $0.001 = 10^x$		−4
8. $5^x = 625$		1
9. $12 = 24 \cdot x^{-1}$		5
10. $6^x \cdot 6^{-3} = 36$		−3
11. $\frac{15^3}{15^x} = 1$		−2
12. $\frac{1}{10{,}000} = 10^x$		6
		4
		−1
		−10

Answer: 1. −2 2. −2 3. 1 4. −5 5. 10 6. 6 7. −3 8. 4 9. 2 10. 5 11. 3 12. −4

Holt Pre-Algebra

What We Are Learning

Rational Numbers and Operations

Vocabulary
These are the math words we are learning:

Denominator The part of a fraction that tells how many equal parts are in the whole.

Least Common Denominator The least common multiple of two or more denominators.

Numerator The part of a fraction that tells how many parts of the whole are being considered.

Rational Number Any number that can be written as a fraction.

Reciprocals Two numbers whose products are equal to one.

Relatively Prime Numbers that have no common factors other than 1.

Dear Family,

Up until this point your child has been working with whole numbers and integers to solve equations and evaluate expressions. In this section your child will be introduced to the set of **rational numbers.** This set of numbers is important because it includes fractions and decimals.

A rational number is any number that can be written as a fraction $\frac{n}{d}$, where n and d are integers and $d \neq 0$. Decimals that repeat or terminate are rational numbers.

Your child will learn to simplify fractions, evaluate expressions using the four basic operations, and solve simple equations and inequalities using numbers from the set of rational numbers.

Writing an answer in simplest form is a skill that will be taught and that your child must master. The goal of simplifying a fraction is to make sure the numerator and the denominator have no common factors other than 1. This is called being **relatively prime.**

Dividing the numerator and the denominator by the same common factor is one simple way of simplifying a fraction.

Simplify.

$\frac{30}{45}$

$30 = 2 \cdot 15$
$45 = 3 \cdot 15$ 15 is a common factor.

$\frac{30}{45} = \frac{30 \div 15}{45 \div 15}$

$\frac{2}{3}$ Divide the numerator and the denominator by 15.

Your child will be asked to write answers in simplest form throughout this section, and in chapters to follow.

Holt Pre-Algebra

Your child will also learn to add, subtract, multiply, and divide rational numbers. When adding or subtracting fractions, there are some basic steps that must be followed.

- If the denominators are the same then ONLY add or subtract the numerators and keep the denominator the same. Write the answer in simplest form.

$$\frac{7}{12} + \frac{3}{12} = \frac{7 + 3}{12} = \frac{10}{12} = \frac{5}{6}$$

- If the denominators are not the same, you must find a common denominator and then rewrite the fractions using the common denominator. Add or subtract as instructed.

$$\frac{3}{4} - \frac{1}{6} = \frac{3}{4}\left(\frac{3}{3}\right) - \frac{1}{6}\left(\frac{2}{2}\right) = \frac{9}{12} - \frac{2}{12} = \frac{7}{12}$$

To multiply fractions, multiply the numerators and multiply the denominators.

To divide fractions, multiply the dividend by the **reciprocal** of the divisor. To find the reciprocal, just flip the numerator and the denominator.

$$\frac{7}{8} \div \frac{2}{5} = \frac{7}{8} \cdot \frac{5}{2} = \frac{7 \cdot 5}{8 \cdot 2} = \frac{35}{16} = 2\frac{3}{16}$$

Discuss with your child the importance of rational numbers in everyday situations. Create problems that will allow your child to practice the skills learned in this section.

Sincerely,

Holt Pre-Algebra

CHAPTER **Family Letter**

3 *Rational Numbers and Operations*

Simplify.

1. $-\dfrac{18}{36}$ | **2.** $\dfrac{15}{35}$ | **3.** $-\dfrac{24}{36}$ | **4.** $\dfrac{25}{40}$

_____ | _____ | _____ | _____

Add or subtract.

5. $\dfrac{8}{12} - \dfrac{7}{12}$ | **6.** $\dfrac{7}{15} + \dfrac{4}{15}$ | **7.** $\dfrac{2}{7} + \dfrac{13}{21}$ | **8.** $\dfrac{-8}{9} + \dfrac{2}{3}$

_____ | _____ | _____ | _____

9. $\dfrac{3}{4} + \dfrac{5}{8}$ | **10.** $\dfrac{5}{6} + \dfrac{3}{4}$ | **11.** $\dfrac{11}{15} - \dfrac{4}{5}$ | **12.** $1\dfrac{1}{4} - 3\dfrac{1}{2}$

_____ | _____ | _____ | _____

Multiply. Write each answer in simplest form.

13. $5\left(\dfrac{1}{7}\right)$ | **14.** $-6\left(1\dfrac{2}{3}\right)$ | **15.** $\dfrac{4}{5}\left(\dfrac{-3}{10}\right)$ | **16.** $\dfrac{7}{8}\left(\dfrac{-3}{5}\right)$

_____ | _____ | _____

Divide. Write each answer in simplest form.

17. $\dfrac{1}{8} \div \dfrac{2}{3}$ | **18.** $\dfrac{3}{4} \div \dfrac{5}{4}$ | **19.** $\dfrac{-3}{4} \div \dfrac{2}{3}$ | **20.** $7\dfrac{1}{2} \div \left(-1\dfrac{1}{4}\right)$

_____ | _____ | _____ | _____

Evaluate each expression for the given value of the variable.

21. $\dfrac{3.1}{x}$ for $x = 0.2$ | **22.** $x \div \dfrac{7}{9}$ for $x = 2\dfrac{4}{9}$ | **23.** $\dfrac{5.4}{x}$ for $x = 1.5$

_____ | _____ | _____

Solve.

24. $x - 18.5 = -32$ | **25.** $\dfrac{x}{11.4} = 3$ | **26.** $x - 3.62 = 7.18$

_____ | _____ | _____

27. $\dfrac{3}{10}x = \dfrac{4}{15}$ | **28.** $x - \dfrac{1}{8} > \dfrac{-3}{4}$ | **29.** $4x \geq 14.4$

_____ | _____ | _____

Answers: 1. $-\dfrac{1}{2}$ **2.** $\dfrac{3}{7}$ **3.** $-\dfrac{2}{3}$ **4.** $\dfrac{5}{8}$ **5.** $\dfrac{1}{12}$ **6.** $\dfrac{11}{15}$ **7.** $\dfrac{19}{21}$ **8.** $-\dfrac{2}{9}$ **9.** $1\dfrac{3}{8}$ **10.** $1\dfrac{7}{12}$ **11.** $-\dfrac{1}{15}$ **12.** $-2\dfrac{1}{4}$ **13.** $\dfrac{5}{7}$ **14.** -10 **15.** $-\dfrac{6}{25}$ **16.** $-\dfrac{21}{40}$ **17.** $\dfrac{3}{16}$ **18.** $\dfrac{3}{5}$ **19.** $-1\dfrac{1}{8}$ **20.** -6 **21.** 15.5 **22.** $3\dfrac{1}{7}$ **23.** 3.6 **24.** $x = -13.5$ **25.** $x = 34.2$ **26.** $x = 10.8$ **27.** $x = \dfrac{8}{9}$ **28.** $x > \dfrac{-5}{8}$ **29.** $x \geq 3.6$

Holt Pre-Algebra

Family Fun
Taskmasters

Directions

- The object of the game is to score the most points.

- Three players or teams play against each other.

- Cut out and shuffle the task cards.

- Player A selects one task card. Player B and Player C complete the task simultaneously.

- If the player is asked to create a problem for the other player, whoever correctly completes the problem first earns the points.

- Player A checks the work. Whoever completes the task/problem correctly earns the points. Answers **MUST** be written in simplest form.

- The players rotate jobs.

- Each player must read the task cards 3 times.

- The team with the most points is the winner!

Task Cards

Create an addition problem for your opponent that involves adding two fractions with unlike denominators. 5 pts	Multiply. Write the answer in simplest form. $\frac{2}{3}\left(\frac{8}{9}\right)$ 2 pts
Create a division problem for your opponent that involves two fractions and whose solution will be negative. 5 pts	Evaluate $4\frac{4}{5}x$ when $x = 20$. 3 pts
Multiply $-9.06\,(0.4)$ 2 pts	Create a multiplication problem that involves a mixed number and a fraction. 5 pts
Solve. Write the answer in simplest form. $x + \frac{7}{8} = \frac{3}{4}$ 3 pts	Solve. Write the answer in simplest form. $\frac{8}{15}t = \frac{7}{9}$ 3 pts
A recipe calls for $\frac{3}{4}$ cups of flour. You need to triple this recipe. How much flour do you need? 2 pts	Subtract. Write the answer in simplest form. $3\frac{7}{12} - 5\frac{1}{9}$ 3 pts

Holt Pre-Algebra

CHAPTER 3 Family Letter
Section B

What We Are Learning

eal Numbers

Vocabulary
These are the math words we are learning:

Density Property A property that states that between any two real numbers there is another real number. This property is also true for rational numbers, but not for whole numbers and integers.

Irrational Number Numbers written as decimals that are not terminating or repeating.

Perfect Squares A number that has integers as its square roots.

Principal Square Root The non-negative square root that appears on the calculator.

Real Numbers The set of numbers that consists of the set of rational numbers and the set of irrational numbers.

Dear Family,

In this section your child will learn about squares and square roots. Although every positive number has two square roots, most of the time you only write the non-negative square root, better known as the **principal square root**.

Your child will also learn to recognize a **perfect square** and use this knowledge to evaluate expressions and estimate the square roots of numbers that are not perfect squares.

Find the two square roots of 81.

$\sqrt{81} = 9$ 9 is a solution, since $9 \cdot 9 = 81$.
$-\sqrt{81} = -9$ -9 is also a solution since $-9 \cdot -9 = 81$.

Your child will need to remember to follow the order of operations when evaluating expressions with exponents.

Evaluate the expression, $4\sqrt{64} + 9$.

$4\sqrt{64} + 9$
  $4(8) + 9$ Evaluate the square root.
  $32 + 9$ Multiply.
  41 Add.

The square root of 55 is between two integers. Name the integers.

$\sqrt{55}$ Think: What perfect squares are close to 55?

$7^2 = 49$ $49 < 55$
$8^2 = 64$ $64 > 55$
$7 < \sqrt{55} < 8$

$\sqrt{55}$ is between 7 and 8.

21

Holt Pre-Algebra

The study of square roots leads to the question of what to call the square roots of numbers that are not perfect squares. These numbers are called **irrational numbers** and are written as decimals that are non-repeating and non-terminating.

There is one last set of numbers that encompasses all the sets of numbers your child has studied. The set of **real numbers** includes the set of whole numbers, integers, rational numbers, and irrational numbers. However, there are some numbers that are not real numbers.

The square root of a negative number or a number divided by zero is NOT considered to be a real number. Your child will learn to identify the set or sets of numbers to which a given number belongs.

Write all names that apply to each number.

$\sqrt{28}$	-91.87	$\dfrac{\sqrt{100}}{5}$
28 is a whole number but not a perfect square.	-91.87 is a terminating decimal.	$\dfrac{\sqrt{100}}{5} = \dfrac{10}{5} = 2$
irrational, real	rational, real	whole, integer, rational, real

The information covered in this section is just an introduction to concepts that will be covered in later courses. Your child will need to have a basic understanding of the various sets of numbers and the characteristics of each set.

Review with your child not only how the sets of numbers build upon each other, but also how the sets are similar, and how they are different.

Sincerely,

Holt Pre-Algebra

CHAPTER 3 Family Letter
Real Numbers

Find the two square roots of each number.

1. 36   **2.** 121   **3.** 64   **4.** 900

_____   _____   _____   _____

Evaluate each expression.

5. $\sqrt{25} - 4$ **6.** $\sqrt{81} - 9$ **7.** $\sqrt{49} + 5$

_____   _____   _____

Each square root is between two integers. Name the integers.

8. $\sqrt{11}$ **9.** $-\sqrt{72}$ **10.** $\sqrt{60}$

_____   _____   _____

Use a calculator to find each value. Round to the nearest tenth.

11. $\sqrt{72}$ **12.** $\sqrt{34.5}$ **13.** $\sqrt{824}$

_____   _____   _____

Write all names that apply to each number.

14. 8   **15.** -11 **16.** $\frac{8}{9}$ **17.** $-\sqrt{3}$

_____   _____   _____   _____

State if the number is rational, irrational, or not a real number.

18. $\frac{-\sqrt{36}}{-6}$ **19.** $-\sqrt{\frac{16}{0}}$ **20.** $\sqrt{5}$ **21.** $\frac{\sqrt{81}}{8}$

_____   _____   _____   _____

Find a real number between each pair of numbers.

22. $4\frac{2}{7}$ and $4\frac{3}{7}$ **23.** $\sqrt{9}$ and 5   **24.** $\frac{-1}{121}$ and 0

_____   _____   _____

Holt Pre-Algebra

Family Fun

Be a Square

Directions

Cut out the cards and shuffle them. Randomly place the cards face down. Taking turns with your opponent, choose two cards and try and match the square root to the perfect square. When you find a pair, keep the cards. If you do not have a match, return each card face down in its original spot. The player with the most matches wins.

$\sqrt{4}$	$\sqrt{9}$	$\sqrt{16}$	$\sqrt{25}$
$\sqrt{36}$	$\sqrt{49}$	$\sqrt{64}$	$\sqrt{81}$
2	3	4	5
6	7	8	9

Holt Pre-Algebra

Family Letter

Section A

What We Are Learning

Collecting and Describing Data

Vocabulary
These are the math words we are learning:

Back-to-Back Stem-and-Leaf Plot A means to compare two sets of data.

Biased Sample A sample that is not based on a good representation of the population.

Box-and-Whisker Plot A way to show the distribution of the data through the use of quartiles and the median.

Mean The sum of the values, divided by the total number of values.

Median If an odd number of values: the middle value. If an even number of values: the average of the two middle values.

Mode The value or values in a data set that occur most often.

Outlier An extreme value that may have an effect on the mean of the data.

Population The entire group being studied.

Quartile The division of a set of data into four equal parts.

Random Sample Every member of the population has an equal chance of being chosen.

Dear Family,

When studying a group or a population it is important to get true and accurate information. While interviewing everyone in a population would achieve the most precise results, this is often not feasible. Usually a **sample** of the **population** is used to represent the views of an entire group.

Identify the population and the sample. Give a reason why the sample could be biased.
A local newspaper sends a survey to 500 local subscribers to find which mayoral candidate people prefer.

Population: People in the local community
Sample: Up to 500 subscribers who take the survey.
Possible Bias: Not all people in the community may subscribe to the local paper.

Your child will learn to organize data so it can be evaluated. A **stem-and-leaf plot** is one method used to organize large amounts of data in a simple, yet precise, manner.

Create a stem-and-leaf plot of the data values.
85, 74, 91, 77, 86, 80, 71, 79, 82, 84, 99, 62

Step 1 Find the least data value and the greatest data value. Since the data values range from 62 to 99, use the tens digits for the **stems** and the ones digits for the **leaves.**

Step 2 List the stems from least to greatest on the plot.

Step 3 List the leaves for each stem from least to greatest. For the number 62, the 6 is the stem and the 2 is the leaf.

Stem	Leaves
6	2
7	1 4 7 9
8	0 2 4 5 6
9	1 9

Holt Pre-Algebra

Range The largest data value minus the smallest data value in a set of data.

Sample Part of the population being surveyed.

Stem-and-Leaf Plot A graph used to organize and display data so that the frequencies can be compared.

Stratified Sample A sampling method that chooses from randomly chosen subgroups.

Systematic Sample A sampling method that follows a rule or formula.

Variability The description of how spread out a data set is.

Once the data is organized, your child will use the measures of central tendency to describe the data.

Mean	The sum of the values divided by the number of values. Otherwise known as the average.
Median	The median is the middle value if there is an odd number of values. If there is an even number of values, the median is the average of the two middle numbers.
Mode	The value or values that occur most often in a set of data. If no value occurs more than once, then there is no mode.

Find the range, mean, median, and mode of the data set.

26, 27, 28, 22, 3, 28, 26, 26, 22, 20

Range

$28 - 3 = 25$ Subtract the least value
The range is 25.   from the greatest value.

Mean

$26 + 27 + 28 + 22 + 3 + 28 + 26 + 26 + 22 + 20 = 228$
                                   Add the values.
$228 \div 10 = 22.8$ Divide the sum by the
The mean is 22.8.   number of items.

Median

3, 20, 22, 22, 26, 26, 26, 27, 28, 28
                                   Arrange the values in order.
                                   Choose the number
The median is 26.   in the middle of the data set.

Mode

The mode is 26.   The value 26 occurs 3 times.

Your child will also learn to identify an outlier. An outlier is an extreme piece of data. For instance, in the data set above, 3 is an outlier. If you remove 3 from the data set the mean becomes 25. Ask your child to explain how the other measures of central tendency are affected when you remove the outlier.

Encourage your child to be aware of the everyday instances of data analysis and the importance it has in our lives.

Sincerely,

Holt Pre-Algebra

Name _____ Date _____ Class _____

Family Letter
Collecting and Describing Data

Identify the population and sample. Give a reason why the sample could be biased.

1. A deli clerk surveys 20 customers to see what kind of cheese is purchased most frequently.

   Population   Sample   Possible Bias

   _____   _____   _____

2. A librarian asks the first four people who return the latest bestseller if they liked the book.

   Population   Sample   Possible Bias

   _____   _____   _____

Identify the sampling method used.

3. The names of all children who are 12 years old and want to play soccer are each written on a piece of paper and placed into a bowl. Each coach selects ten names, one at a time, to make a team.

   _____

List the data values in the stem-and-leaf plot.

4.
1	1 2
2	3 4 6
3	2 8

   _____

5.
4	0 1 2
5	1 5 8
6	0 1 1 5
7	0 3 5 5 8

   _____

Find the mean, median, and mode of each data set.

6. 15, 40, 22, 46, 7, 20, 22, 12

   mean   median   mode

   _____   _____   _____

7. 106, 46, 93, 98, 92, 95, 100

   mean   median   mode

   _____   _____   _____

Find the range and the first and third quartiles for each data set.

8. 14, 8, 25, 26, 11, 16, 23, 15, 40

   range   first quartile   third quartile

   _____   _____   _____

9. 94, 85, 76, 98, 82, 87, 78

   range   first quartile   third quartile

   _____   _____   _____

Answers: 1. Population: all store customers; Sample: 20 customers; Possible Bias: Not all customers buy or like cheese 2. Population: All people who checked out books; Sample: 4 people who returned bestseller first; Possible Bias: Not all readers may have finished the book; person bringing back the book did not read it. 3. random 4. 11, 12, 23, 24, 26, 32, 38 5. 40, 41, 42, 51, 55, 58, 60, 61, 61, 65, 70, 73, 75, 78 6. mean: 23; median: 21; mode 22 7. mean: 90; median: 95; mode: no mode 8. range: 32; first quartile: 12.5; third quartile: 25.5 9. range: 22; first quartile: 78; third quartile: 94

Holt Pre-Algebra

Name _____ Date _____ Class _____

Directions
Use the math words from this section to solve the puzzle.

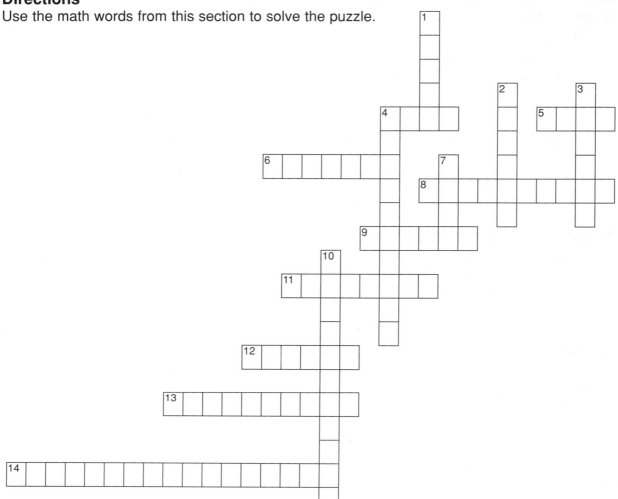

Across
4. The _____ is the first value in this type of plot.
5. Average
6. An extreme value that affects the mean.
8. The group being studied.
9. This type of sample does not represent the whole population
11. The division of data into four equal parts.
12. The middle number in a set of data
13. Sampling method example - Every third child will get a letter home.
14. Shows the distribution of data using the median and quartiles

Down
1. The difference between the largest value and the smallest value.
2. The part of the population being studied.
3. A sampling method where anyone in the population can be chosen.
4. A type of sample that chooses from subgroups.
7. In the set 3, 7, 3, 4, 6, 8, 5, 3, 8, the _____ is 3.
10. Description of a data spread.

Answers: Across: 4. stem 5. mean 6. outlier 8. population 9. biased 11. quartile 12. median 13. systematic 14. box and whisker plot Down: 1. range 2. sample 3. random 4. stratified 7. mode 10. variability

Holt Pre-Algebra

What We Are Learning

Displaying Data

Vocabulary
These are the math words we are learning:

Bar Graph A way to display data that can be grouped into categories.

Correlation Describes the type of relationship between two data sets.

Frequency Table A way to organize data in terms of the number of times each value occurs.

Histogram A type of bar graph where the data is grouped in intervals

Line Graph A type of graph often used to show trends or make estimates for values between data points.

Line of Best Fit The line that comes closest to all the points on a scatter plot.

Scatter Plot Shows the relationship between two sets of data.

Dear Family,

In this section your child will be interpreting and constructing many different types of graphs used to display data.

Bar graphs are helpful when you want to display data that can be put into categories.

Frequency tables help you organize data so you can tell how many items fall into a particular category.

A **histogram** is a type of bar graph that shows the frequency of data within equal intervals. Shown below is an example of how to construct a histogram.

The frequency table shows the average number of hours per day in the summer that teenagers spend sleeping. Use this data to make a histogram.

Number of hours per day spent sleeping	Frequency
5–6	5
7–8	8
9–10	10
11–12	12

Step 1 Choose an appropriate scale and interval for the vertical axis. The greatest value for this axis should be at least as great as the greatest frequency.

Step 2 Draw a bar for each interval. The height of the bar is the frequency for that interval. Bars must touch, not overlap.

Step 3 Label both axes. Give the graph a title.

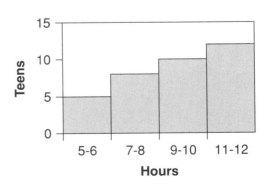

Sleeping Teens

Holt Pre-Algebra

As your child learns to create different types of graphs, it is critical that they learn to distinguish an accurate portrayal of data versus a misleading one. Being able to recognize a misleading graph or statistic is an essential tool for a smart consumer.

Explain why the graph is misleading.

Preferred Brands

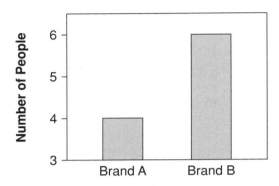

By just looking at the graph it appears that twice as many people prefer Brand B than Brand A. However, if you look closely, notice that the scale does not start at zero. The scale starts at 3. By reading the graph you can learn that 4 people prefer Brand A and 6 people prefer Brand B.

The graph is misleading because the difference between which brand people prefer is not as big as the graph portrays.

Encourage your child to be aware of the day-to-day examples of how data is presented, especially in the news and newspaper. Discuss these displays with your child to help him or her gain understanding of the concepts presented in this section.

Sincerely,

Holt Pre-Algebra

Name _____ Date _____ Class _____

Family Letter
Displaying Data

1. Organize the data into a frequency table and make a bar graph.

| 25 | 24 | 22 | 26 | 23 | 26 | 29 |
| 22 | 26 | 26 | 23 | 28 | 25 | 27 |

Interval	Frequency
22–23	
24–25	
26–27	
28–29	

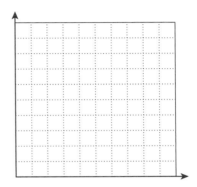

Explain why the statistic is misleading.

2. A school reporter asked 100 students what their favorite subject was. Of the 30 that responded, 14 said Math, 8 said Science, 5 said History, and 3 said Physical Education. The reporter wrote, *"Half of all the students pick Math as their favorite subject."*

_____

3. A group of students had babysitting jobs over the weekend. Use the given data to make a scatter plot.

Name	Total Hours Worked	Amount Earned
Sheri	5	$22.50
Jordan	8	$36.00
Lydia	11	$49.50
Alexis	6	$27.00

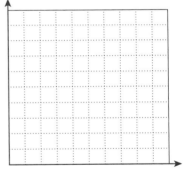

4. Use the data from the scatter plot to predict the number of hours Sam would have to work to make $30.

_____

Do the sets of data have a positive, negative, or no correlation?

5. The weight of a baby and the month that it is born.

6. The amount of free time you have and the number of sports that you play.

_____ _____

Answers: 1. Table: 4 marks, 3 marks, 5 marks, 2 marks; Bars on the graph should measure accordingly. **2.** The sample response was too small. Only 30% responded and less than 50% stated Math was their favorite subject. **3.** Points should be plotted at (5, 22.50), (8, 36), (11, 49.50) and (6, 27); *x*-axis is hours worked, *y*-axis is amount earned **4.** about 7 hours **5.** no correlation **6.** negative correlation

Holt Pre-Algebra

CHAPTER	**Family Fun**
4	*Graphing for Greatness*

It is said that there is a statistic for just about everything in the game of baseball. Choose a player and a baseball statistic, like homeruns, strikeouts, RBI's, or stolen bases. Make a line graph on the graph below using your player's statistics over a given period of time. Be sure to label the axes and title your graph. Then answer the questions below.

For the non-baseball fan, research the statistics of another athlete, such as a tennis player, golfer, swimmer, or gymnast.

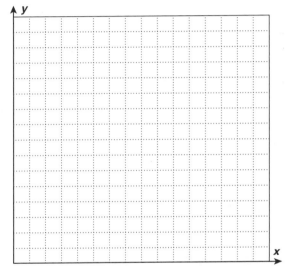

1. Who is your athlete? What team was he involved with, if any?

   _____

2. Which statistic did you investigate? _____

3. Which seasons/years were the most successful for your athlete?

   _____

4. Which seasons/years did your athlete seem to struggle?

   _____

5. How long did your player play his/her sport? _____

6. What, if anything, affected your player's performance? (i.e. age, illness, experience)

   _____

7. How do you think this information helped the opposing team compete against your player?

   _____

Holt Pre-Algebra

Family Letter
Section A

What We Are Learning

Plane Figures

Vocabulary
These are the math words we are learning:

Angle Formed by two rays with a common endpoint called the vertex.

Congruent Figures with the same size and the same shape.

Parallel Lines Two lines in a plane that never meet.

Parallelogram A quadrilateral with 2 pairs of parallel sides.

Perpendicular Lines Two lines that intersect at 90° angles.

Polygon A closed plane figure formed by three or more segments.

Ray A part of a line that starts at one point and extends forever in one direction.

Rectangle A quadrilateral with 4 right angles.

Regular Polygon A polygon in which the sides and the angles have equal measures.

Rhombus A quadrilateral with 4 congruent sides.

Rise The vertical change in a line.

Run The horizontal change in a line.

Segment A part of a line between two points.

Dear Family,

In this section, your child will begin the study of geometry. During this process, your child will learn to classify and name many geometrical figures. First, your child will review the foundation of geometry, which includes the definitions and the identification of points, lines, planes, segments, and rays.

Naming Points, Lines, Planes, Segments, and Rays

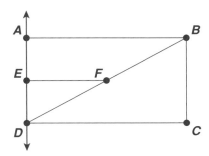

Name six points in the figure.
point *A*, point *B*, point *C*, point *D*, point *E*, and point *F*

Name two lines in the figure.
$\overleftrightarrow{DA}$, $\overleftrightarrow{EA}$ Any 2 points on the line can be used.

Name a plane in the figure.
plane *AEF* or any 3 points in the plane that form a triangle

Name four segments in the figure.
$\overline{AB}$, $\overline{CD}$, $\overline{EF}$, $\overline{BC}$

Name two rays in the figure.
$\overrightarrow{AE}$, $\overrightarrow{DE}$

Angles are very important when studying geometry. Special angles have special names and relationships. Knowing these relationships will allow your child to find unknown angle measurements.

Acute: Angles that measure less than 90°.

Obtuse: Angles that measure more than 90° but less than 180°.

Right: Angles that measure exactly 90°.

Complementary: Angles whose measures add to 90°.

Supplementary: Angles whose measures add to 180°.

Holt Pre-Algebra

Family Letter

Section A, continued

Slope A number that describes the steepness of a line.

Square A quadrilateral with 4 congruent sides and 4 right angles.

Trapezoid A quadrilateral with exactly one pair of parallel sides.

Transversal A line that intersects any two or more other lines.

Polygons are also an important part of geometry. One special polygon is a triangle. Your child will learn how to classify triangles by the angles or side lengths.

Triangle	Description	Picture
Acute Triangle	A triangle with 3 acute angles.	
Right Triangle	A triangle with 1 right angle.	
Obtuse Triangle	A triangle with 1 obtuse angle.	
Equilateral Triangle	A triangle with 3 congruent sides and 3 congruent angles.	
Isosceles Triangle	A triangle with 2 congruent sides and 2 congruent angles.	
Scalene Triangle	A triangle with no congruent sides or congruent angles.	

Your child will use the **Triangle Sum Theorem** to find unknown angle measures in a triangle. The theorem states that the sum of the angles of any triangle measures 180°.

Find the angle measures in the isosceles triangle.

$68° + k° + k° = 180°$ — Triangle Sum Theorem
$68° + 2k° = 180°$ — Combine like terms.
$\underline{-68° \quad -68°}$ — Subtract 68° from both sides.
$2k° = 112°$
$\dfrac{2k°}{2} = \dfrac{112°}{2}$ — Divide both sides by 2.
$k° = 56°$

The triangle has angle measures of 56°, 56° and 68°.

By understanding these basic concepts of triangles, your child will be well on their way to solving problems involving other types of polygons.

Sincerely,

Holt Pre-Algebra

Family Letter

CHAPTER 5

Plane Figures

1. Name four points in the figure.

   _____

2. Name two lines in the figure.

   _____

3. Name three rays in the figure.

   _____

4. Name a right angle in the figure.

   _____

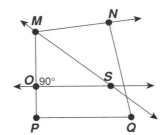

Find the value of each variable.

5.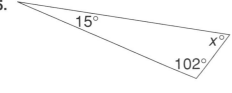

   _____

6.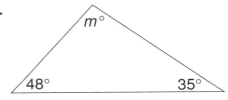

   _____

Write all the names that apply to each figure.

7.

   _____

8.

   _____

9.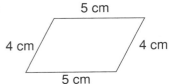

   _____

Find the sum of the angle measures in each figure.

10. regular octagon

11. irregular hexagon

12. irregular pentagon

_____

_____

_____

Answers: 1. Points M, N, O, P, Q, S **2.** $\overleftrightarrow{MS}$, $\overleftrightarrow{OS}$ **3.** $\overrightarrow{MN}$, $\overrightarrow{SM}$, $\overrightarrow{SO}$, $\overrightarrow{MS}$, $\overrightarrow{OS}$ **4.** ∠MOS, ∠SOP **5.** $63°$
6. $97°$ **7.** quadrilateral, parallelogram, rectangle, square **8.** quadrilateral, trapezoid
9. quadrilateral, parallelogram **10.** $1080°$ **11.** $720°$ **12.** $540°$

Holt Pre-Algebra

Name _____ Date _____ Class _____

Family Fun
Scavenger Hunt

Directions

In teams of 2, you have 15 minutes to tour your home and search for as many unusual shapes as possible. The shapes may be inside or outside. Your team must have at least one object per shape. If the shape has 3 or 4 sides, it is worth 2 points. If a shape has more than 4 sides, it is worth 5 points. Record each object your team finds under the name of each shape. The team with the most points at the end of the time limit wins!

square	rectangle	trapezoid	octagon	parallelogram	right triangle
pentagon	hexagon	isosceles triangle	rhombus	equilateral triangle	scalene triangle

square	rectangle	trapezoid	octagon	parallelogram	right triangle
pentagon	hexagon	isosceles triangle	rhombus	equilateral triangle	scalene triangle

Holt Pre-Algebra

What We Are Learning

Patterns in Goemetry

Vocabulary
These are the math words we are learning:

Center of Rotation The point about which a figure is rotated.

Correspondence A way of matching up two sets of objects.

Image The resulting figure after a translation, rotation, or reflection.

Line of Symmetry The line that creates line symmetry.

Line Symmetry Occurs when a line is drawn through a figure and two sides are mirror images of each other.

Reflection Flips a figure across a line to create a mirror image.

Rotation Turns a figure around a point.

Rotational Symmetry Occurs when a figure is rotated less than 360° around a central point so it coincides with itself.

Regular Tessellation A regular polygon is repeated to fill a plane.

Semi-regular Tessellation Two or more regular polygons are repeated to fill the plane and the vertices are identical.

Dear Family,

In this section, your child will begin to apply geometrical properties to solve problems. One such property addresses congruent figures. As your child previously learned, if two polygons are congruent, then all of their corresponding sides and angles are also congruent.

Your child will write congruence statements for pairs of polygons by writing the second polygon in order of correspondence, or matching up the corresponding vertices between the two polygons.

Write a congruence statement for this pair of polygons.

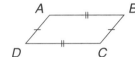 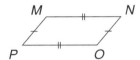

$\angle A \cong \angle M$ so $\angle A$ corresponds to $\angle M$.

$\angle B \cong \angle N$ so $\angle B$ corresponds to $\angle N$.

$\angle C \cong \angle O$ so $\angle C$ corresponds to $\angle O$.

$\angle D \cong \angle P$ so $\angle D$ corresponds to $\angle P$.

The congruence statement is parallelogram $ABCD \cong$ parallelogram $MNOP$.

Note how the vertices in the first polygon are written in order around the polygon starting at any vertex. The vertices in the second polygon, therefore, have to be written in the same order.

If you know polygons are congruent, you can find an unknown value in the polygon.

In the figure, triangle $RST \cong$ triangle UVW. Find m.

$m - 7 = 25$ $ST \cong VW$

$$m - 7 = 25$$
$$\underline{+7 \quad +7}$$ Add 7 to both sides.
$$m = 32$$

The value of m is 32.

Holt Pre-Algebra

Tessellation A repeating pattern of plane figures that completely covers a plane without gaps or overlaps.

Transformation Translating, reflecting, or rotating an object.

Translation Slides a figure along a line without turning.

Another geometrical concept your child will be learning is **transformations.** Transformations include rotations, reflections, and translations of congruent figures.

If you only move a figure along a line it is called a **translation.** If you turn a figure around a point, it is called a **rotation.** If you flip a figure across a line to create a mirror image, it is called a **reflection.**

Identify each as a translation, rotation, reflection, or none.

A.
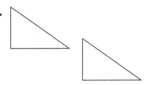

This is called a translation.

B.

This is called a reflection.

C.

This is called a rotation.

D.

This is not a transformation.

Symmetry is another geometrical concept that your child will learn to identify in figures. An object is symmetrical when a line of symmetry is drawn through it, and the resulting sides are mirror images of each other. There are many instances of symmetry in nature. Have your child give you examples of objects in nature that are symmetrical.

Complete the figure. The dashed line is the line of symmetry.

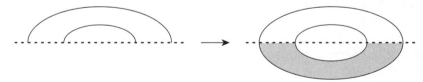

The shaded region is symmetrical to the non-shaded region.

Your child will have a solid background in geometry as the concepts in this section are explored. Discuss with your child the many applications geometry has in our lives.

Sincerely,

Holt Pre-Algebra

CHAPTER 5 **Family Letter**
Patterns in Geometry

Write a congruence statement for each pair of polygons.

1.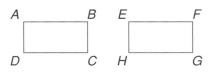

2.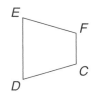

_____   _____

In the figure, triangle *DEF* ≅ triangle *PQR*. Find the value of the variable.

3. Find *r*. _____   4. Find *s*. _____

5. Find *t*. _____   6. Find *u*. _____

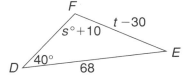

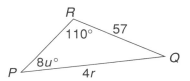

Identify each as a translation, rotation, reflection, or none of these.

7.   8.

_____   _____

9.    10.

_____   _____

Complete the figure. The dashed line is the line of symmetry.

11.   12.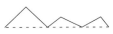

_____   _____

Answers: 1. Rectangle *ABCD* ≅ *EFGH* **2.** Trapezoid *CDEF* ≅ *HIJK* **3.** *r* = 17 **4.** *s* = 100° **5.** *t* = 87 **6.** *u* = 5° **7.** rotation **8.** translation **9.** none of these **10.** reflection **11.** ⊡ **12.** ⋀⋀

Holt Pre-Algebra

Family Fun

Transforming Words

What happens to some words or letters when they undergo a transformation? Which capital letters are symmetrical? Which words can be reflected and still stay the same? Which words change meanings when you rotate them?

Draw a line of symmetry on the capital letters that are symmetrical.

A B C D E F G H I J K L M

N O P R S T U V W X Y Z

Find 4 words which can be reflected and not lose their meaning.

_____

_____

_____

_____

Find 4 words that can be rotated 180° and still be a word.

_____

_____

_____

_____

Print your name horizontally. Make a horizontal reflection of your name.

_____

Print your name vertically. Make a vertical reflection of your name.

_____

Create a secret message in which the only way it can be understood is if it undergoes a transformation.

_____

Answers: A B C D E H I K M O T U V W X Y sample: MOM WOW TAT MUM
sample: MOM/WOW, ON/NO, MOW/WOM, NOON/NOON

What We Are Learning

Perimeter and Area

Vocabulary
These are the math words we are learning:

Area The number of square units in a figure.

Circle A set of points in a plane that are a fixed distance from a given point called the center.

Circumference The distance around the circle.

Diameter A segment that passes through the center and connects any two points on the circle.

Hypotenuse The side on a right triangle across from the right angle.

Leg One of the two sides that make up the right angle in a right triangle.

Perimeter The distance around the outside of a figure.

Pythagorean Theorem In any right triangle, the sum of the squares of the legs is equal to the square of the length of the hypotenuse.

Radius A segment that connects the center to any point on the circle.

Dear Family,

Your child will be learning to find the perimeter and the area of polygons. **Perimeter** is the distance around a polygon. **Area** is defined as the number of square units in a figure.

In rectangles and parallelograms, a simple formula is used to calculate the area: *Area* equals the *base length* times the *height length* or $A = bh$.

Find the area and the perimeter of the figure.

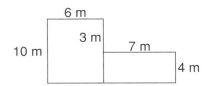

The length of the side that is not labeled is the same as the length of the opposite side.

$p = 10 + 6 + 3 + 7 + 4 + 7 + 6 = 43$ m

$A = (10 \cdot 6) + (7 \cdot 4)$ Add the areas together.

$A = 60 + 28$

$A = 88$ m^2

The perimeter is 43 m, and the area is 88 m^2.

While the process of finding perimeter stays the same for other polygons, your child will need to learn the area formulas for other polygons, like a triangle and a trapezoid.

	Area Formulas for Triangles and Trapezoids	
	Words	**Formula**
Triangle	Area equals one-half the base times the height.	$A = \frac{1}{2} bh$
Trapezoid	Area equals one-half the sum of the bases times the height.	$A = \frac{1}{2} (b_1 + b_2)h$

These formulas are necessary for your child to remember so he or she can calculate the area of a polygon that can be separated into parallelograms, triangles, and trapezoids.

Holt Pre-Algebra

Your child will learn another important formula called the Pythagorean Theorem. This theorem relates the length of the sides, or **legs,** of a right triangle to the length of the **hypotenuse.** The hypotenuse is the side of the triangle directly across from the right angle.

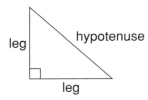

The **Pythagorean Theorem** states that the sum of the squares of the lengths of the two legs is equal to the square of the length of the hypotenuse. Sometimes, the formula is a little easier to understand.

$$a^2 + b^2 = c^2$$
$$\downarrow \qquad \downarrow \qquad \downarrow$$

legs   hypotenuse

Find the length of the hypotenuse.

$a^2 + b^2 = c^2$	Pythagorean Theorem
$5^2 + 12^2 = c^2$	Substitute values.
$25 + 144 = c^2$	Simplify powers.
$169 = c^2$	
$\sqrt{169} = \sqrt{c^2}$	Solve for c: $c = \sqrt{c^2}$
$13 = c$	

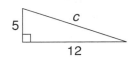

Finally, in this section, your child will use formulas to find the area and circumference of a circle. The **circumference** is the distance around the circle.

The formula for the circumference of a circle is $2\pi r$ or πd. The formula for the area of a circle is πr^2.

The **diameter,** d, is a line segment that connects two points on the circle, but must pass through center of the circle. The **radius,** r, is a line segment that connects a point on the circle to the center. The length of the radius is always one-half the length of the diameter.

Your child will need a firm grasp of the concepts in this lesson, as they provide the foundation for the next lesson.

Sincerely,

Holt Pre-Algebra

Name _____ Date _____ Class _____

Graph and find the area of each figure with the given vertices.

1. (−4, 0), (3, 0), (3, −4), (−4, −4)

2. (1, 4), (7, 4), (4, 0), (−2, 0)

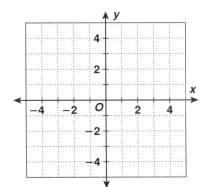

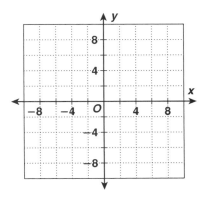

Find the perimeter and area of the figure.

3. Perimeter   Area

_____   _____

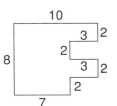

Find the perimeter of each figure.

4.

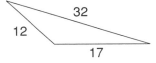

5.

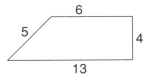

Find the length of the hypotenuse in each triangle.

6.

7.

8. Find the circumference of a circle with radius 8 in., both in terms of π and to the nearest tenth of a unit. Use 3.14 for π.

_____

9. Find the area of a circle with diameter 34 cm, both in terms of π and to the nearest tenth of a unit. Use 3.14 for π.

_____

Answers: 1. rectangle with area of 28 sq. units. 2. parallelogram with area of 24 sq. units 3. $p = 42$ units; $A = 68$ sq. units 4. $p = 61$ units 5. $p = 28$ units 6. $c = 34$ 7. $c = 19.8$ 8. $16\,\pi$ in. ≈ 50.2 in. 9. $289\,\pi$ cm$^2 \approx 907.5$ cm^2

Holt Pre-Algebra

Name _____ Date _____ Class _____

Family Fun
Area versus Perimeter

Materials

Paper   Game cards
Pencil   Number cubes

Directions

• Cut out the cards, shuffle them, and place them face down.

• Roll the number cube. If the number is even, you need to calculate the area of the figure. If the number is odd, you need to calculate the perimeter (or circumference) for that round.

• Each player draws a card and completes the necessary calculation. If the calculation is correct, you receive 10 points. If the calculation is incorrect, you lose 5 points.

• The player with the most points after all the cards have been played is the winner.

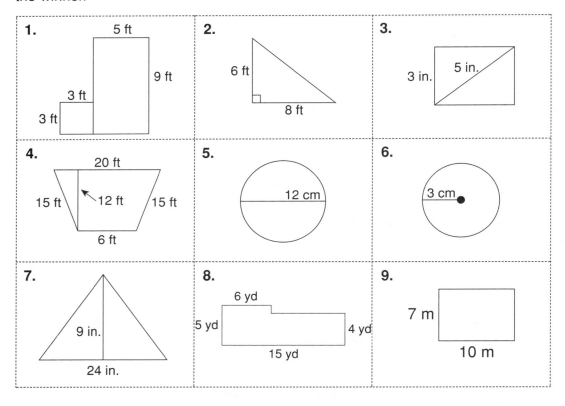

Answers: 1. $p = 34$ ft, $A = 54$ ft^2 2. $p = 24$ ft, $A = 24$ ft^2 3. $p = 12$ in., $A = 12$ in.^{2} 4. $p = 56$ ft, $A = 156$ ft^2 5. $C = 12\pi$ cm, $A = 36\pi$ cm^2 6. $C = 6\pi$ cm, $A = 9\pi$ cm^2 7. $p = 54$ in., $A = 108$ in.^{2} 8. $p = 40$ yd, $A = 66$ yd^2 9. $p = 34$ m, $A = 70$ m^2

Holt Pre-Algebra

What We Are Learning

Three-Dimensional Geometry

Vocabulary
These are the math words we are learning:

Altitude A segment from the vertex perpendicular to the base.

Cone A figure with a circular base.

Cylinder A three-dimensional figure with two circular, parallel, and congruent bases.

Edge The intersection of two faces.

Face A flat surface.

Great Circle The edge of a hemisphere.

Hemisphere Two halves of a sphere.

Lateral Face Parallelograms that connect the bases.

Lateral Surface The curved surface of a cylinder.

Perspective A technique used to make drawings of three-dimensional objects appearing to have depth and distance.

Prism A three-dimensional figure named for the shape of its bases. The two bases are congruent polygons. The rest of the faces are parallelograms.

Dear Family,

Now that your child has a strong foundation in basic two-dimensional geometry, your child will learn to draw and identify three-dimensional figures. Three-dimensional figures have faces, edges, and vertices. The faces are the flat surfaces of the figure; the edges are where two faces intersect; the vertices are where three or more edges intersect.

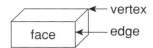

Your child will draw three-dimensional figures by using **perspective,** which is a technique used to make a figure or drawing appear to have depth and distance. In a one-point perspective there is one **vanishing point.** The vanishing point is the point to which the lines are drawn.

Sketch a one-point perspective drawing of a rectangular box.

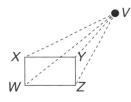

Step 1 Draw a rectangle. This will be the front face. Label the vertices W through Z.

Step 2 Mark a vanishing point, V, somewhere above the rectangle and draw a dashed line from each vertex to V.

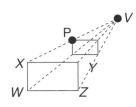

Step 3 Choose a point P on $\overline{XV}$. Lightly draw a smaller rectangle that has P as one of its vertices.

Step 4 Connect the vertices of the two rectangles along the dashed lines.

Step 5 Darken the visible edges, and draw dashed segments for the hidden edges. Erase the vanishing point and the lines connecting it to the vertices.

Holt Pre-Algebra

Pyramid A figure named for the shape of its base. The base is always a polygon and all the other faces are triangles.

Rectangular Prism A prism with a rectangle for its bases.

Regular Pyramid A pyramid with a regular polygon for a base and congruent lateral faces.

Right Cone A cone with a right angle.

Slant Height Measured along the lateral surface of a cone or pyramid.

Sphere A set of points in three dimensions that is a fixed distance from the center.

Surface Area The sum of the areas of all surfaces of a figure.

Triangular Prism A prism with a triangle for its bases.

Vanishing Point In a perspective drawing, the point at which parallel segments of an object meet.

Vertex The intersection of three or more edges.

Your child will also be learning how to calculate the volume and surface area of three-dimensional figures. Use the chart below to help your child learn the volume and surface area formulas for common three-dimensional figures.

Solid Figure	Words	Volume Formulas
Prism	the **area** of the base times the height	$V = Bh$
Cylinder	the **area** of the base times the height	$V = Bh$ $= (\pi r^2)h$
Pyramid	one-third the **area** of the base times the height.	$V = \frac{1}{3}Bh$
Cone	one-third of the area of the **circular base** times the height.	$V = \frac{1}{3}Bh$ $V = \frac{1}{3}\pi r^2 h$
Sphere	$\frac{4}{3}\pi$ times the cube of the radius	$V = \left(\frac{4}{3}\right)\pi r^3$

Solid Figure	Words	Surface Area Formulas
Prism	the area of the bases plus the area of the lateral faces	$S = 2B + F$ $= 2B + ph$
Cylinder	the area of the bases plus the area of the lateral surface area	$S = 2B + L$ $= 2\pi r^2 + 2\pi rh$
Pyramid	the area of the base plus the area of the lateral faces	$S = B + F$ $= B + \frac{1}{2}pl$
Cone	the area of the base plus the area of the lateral surface area	$S = B + L$ $= \pi r^2 + \pi rl$
Sphere	4π times the square of the radius.	$S = 4\pi r^2$

These formulas can be a bit daunting if you do not take the time to practice using the formulas. Review these formulas by quizzing your child.

Sincerely,

Holt Pre-Algebra

Name _____ Date _____ Class _____

Use the one-point perspective drawing for Exercises 1–3.

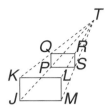

1. Which point is the vanishing point? _____

2. Which face is the front face? _____

3. Which face is the back face? _____

Find the volume of each figure to the nearest tenth of a unit.

4.
8 in.
15 in.

5.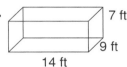
7 ft
9 ft
14 ft

6.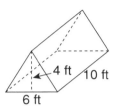
4 ft
10 ft
6 ft

_____   _____   _____

Find the volume of each figure.

7.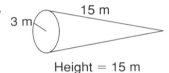
15 m
3 m
Height = 15 m

8.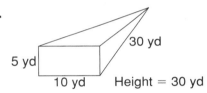
30 yd
5 yd
10 yd
Height = 30 yd

_____   _____

Find the surface area of each figure.

9.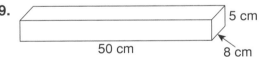
5 cm
50 cm
8 cm

10.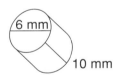
6 mm
10 mm

_____   _____

Find the volume of each sphere, both in terms of π and to the nearest tenth of a unit.

11.
Radius = 5 cm

12.
Diameter = 8 ft

_____   _____

Answers: 1. T 2. *PQRS* 3. *JKLM* 4. 3014.4 in.³ 5. 882 ft³ 6. 120 ft³ 7. 45 π m³ 8. 500 yd³ 9. 1380 cm² 10. 245 mm² 11. 166.7π cm³ or 523.3 cm³ 12. 85.3π ft³ or 267.9 ft³

Holt Pre-Algebra

Family Fun
3-D Riddles

Try and figure out which solid figure is best described by the riddle. Name the solid figure and draw a picture of it.

1. When you *roll* me out, I look like a rectangle.
What am I?

_____

2. My faces have three edges, but my base is square.
What am I?

_____

3. Although my base is round, I always get to the point.
What am I?

_____

4. I made the perfect container for stop signs.
What am I?

_____

5. I am just square all over!
What am I?

_____

6. I am covered with points, but am not sharp.
What am I?

_____

7. Some people call me great, but that is only half true.
What am I?

_____

Answers: 1. cylinder 2. Pyramid 3. cone 4. octagonal prism 5. cube 6. sphere 7. hemisphere

Holt Pre-Algebra

What We Are Learning

Ratios, Rates, and Proportions

Vocabulary

These are the math words we are learning:

Conversion Factor A fraction whose numerator and denominator represent the same quantity but use different units.

Cross Product In a proportion, the product of a numerator on one side with the denominator on the other.

Equivalent Ratio Ratios that name the same comparison.

Proportion An equation that states that two ratios are equivalent.

Rate A ratio that compares two quantities measured in different units.

Ratio A comparison of two quantities by division.

Similar Figures that have the same shape, but not necessarily the same size.

Unit Price A unit rate used to compare prices.

Unit Rate A rate in which the second quantity in the comparison is one unit.

Dear Family,

Your child will be learning to find equivalent ratios in order to create proportions. A **ratio** is simply a comparison of two items by division. Ratios are often expressed as fractions.

Your child will use ratios that name the same comparison as another ratio. These ratios are called **equivalent ratios.** One way your child can determine if two ratios are equivalent, or in **proportion,** is if the ratios can be simplified to the same value.

Simplify to tell whether the ratios form a proportion.

$\frac{10}{15}$ and $\frac{24}{40}$

$$\frac{10}{15} = \frac{10 \div 5}{15 \div 5} = \frac{2}{3} \qquad \frac{24}{40} = \frac{24 \div 8}{40 \div 8} = \frac{3}{5}$$

Since, $\frac{2}{3} \neq \frac{3}{5}$, the ratios are not in proportion.

Another way to determine if two ratios are proportional is to find the cross products of two ratios. If the **cross products** of the ratios are equal, the ratios are in proportion. If the cross products are not equal, the ratios are NOT in proportion.

Tell whether the ratios are proportional.

$\frac{4}{12} \overset{?}{=} \frac{3}{9}$

$\frac{4}{12} \diagdown\!\!\!\diagup \frac{3}{9}$ Find the cross products.

$36 = 36$

Since the cross products are equal, the ratios are proportional.

Your child will apply the properties of proportions to help with problems that involve ratios. When one of the values of a proportion is unknown, your child will learn to solve for that missing value by using cross products.

Holt Pre-Algebra

Your child will also use proportions and conversion factors to solve problems involving **rates.**

You may use some of these common conversions when your child needs to convert units in order to find a solution.

Measure	Customary System	Metric System
Length and Distance	12 in. = 1 ft 3 ft = 1 yd 5280 ft = 1 mi	10 mm = 1 cm 100 cm = 1 m 1000 m = 1 km
Volume and Capacity	2 cups = 1 pt 2 pints = 1 qt 4 qt = 1 gal	1000 mL = 1 L
Weight and Mass	16 oz = 1 lb 2000 lb = 1 ton	1000 mg = 1 g 1000 g = 1 kg

Mary is filling up the sand box with 8 bags of sand. Each bag weighs 3 lb. Use conversion factors to find how many ounces of sand are in each bag.

The problem gives the ratio 3 lb *to* 1 bag and asks for the answer in *ounces* per bag.

$\dfrac{3 \text{ lb}}{1 \text{ bag}} \cdot \dfrac{16 \text{ oz}}{1 \text{ lb}}$ Multiply by the conversion factor.

$= \dfrac{3 \cdot 16 \text{ oz}}{1 \text{ bag}}$ Cancel the lb units.

$= 48$ oz per bag

Have your child solve proportions that involve real life information. This will allow your child to see the application of this concept outside the classroom.

Sincerely,

Holt Pre-Algebra

CHAPTER 7	# Family Letter

Ratios, Rates, and Proportions

Find two ratios that are equivalent to each given ratio.

1. $\frac{3}{8}$

2. $\frac{9}{4}$

3. $\frac{20}{30}$

_____

Simplify to tell whether the ratios form a proportion.

4. $\frac{12}{15}$ and $\frac{28}{35}$

5. $\frac{12}{30}$ and $\frac{30}{75}$

6. $\frac{45}{63}$ and $\frac{36}{96}$

_____

Determine the better buy.

7. 6 cans of soup for $3.39
or 4 cans of soup for $2.29

8. 28-oz bottle of juice for $2.29
or 64-oz bottle of juice for $5.59

_____

Find the appropriate factor for each conversion.

9. km to m

10. feet to yards

11. quarts to pints

_____

12. A truck traveled 550 ft down a road in 11 seconds. How many miles per hour was the truck traveling?

_____

Tell whether each pair of ratios is proportional.

13. $\frac{15}{24}$ and $\frac{5}{8}$

14. $\frac{9}{7}$ and $\frac{11}{19}$

15. $\frac{14}{16}$ and $\frac{35}{40}$

_____

Solve each proportion.

16. $\frac{m}{9} = \frac{16}{12}$

17. $\frac{12}{7} = \frac{42}{x}$

18. $\frac{3}{5} = \frac{d}{60}$

_____

Answers: For 1–3 possible answers: 1. $\frac{6}{16}$, $\frac{12}{32}$ 2. $\frac{18}{8}$, $\frac{45}{20}$ 3. $\frac{2}{3}$, $\frac{12}{18}$ 4. The ratios are in proportion. 5. The ratios are in proportion. 6. The ratios are not in proportion. 7. 6 cans for $3.39 8. 28 oz for $2.29 9. $\frac{1000 \text{ m}}{1 \text{km}}$ 10. $\frac{1 \text{ yd}}{3 \text{ ft}}$ 11. $\frac{2 \text{ pt}}{1 \text{ qt}}$ 12. 34 mi/h 13. yes 14. no 15. yes 16. $m = 12$ 17. $x = 24.5$ 18. $d = 36$

Holt Pre-Algebra

Family Fun
Proportional Pictures

Materials
Red and blue markers

Directions
Solve the proportions. If the last digit of the unknown value is even, color that particular area blue. If the last digit of the answer is odd, color that particular area red.

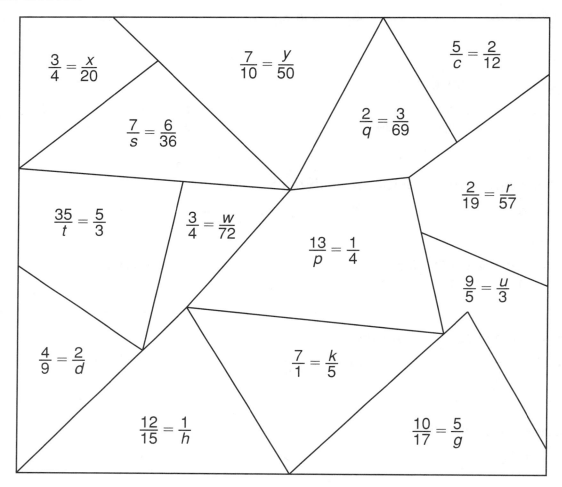

Holt Pre-Algebra

Family Letter

CHAPTER 7 Section B

What We Are Learning

Similarity and Scale

Vocabulary
These are the math words we are learning:

Capacity   The amount a container can hold when filled.

Center of Dilation   The point of intersection of the lines extending from all the angles in a dilation.

Dilation   A transformation that changes the size, but not the shape of a figure.

Enlargement   An increase in size of all dimensions.

Reduction   A decrease in size of all dimensions.

Similar   Figures that have the same shape, but not necessarily the same size.

Scale   The ratio between two sets of measurements.

Scale Drawing   A drawing that uses a scale to make an object smaller than or larger than the real object.

Scale Factor   The ratio used to enlarge or reduce similar figures.

Scale Model   A three-dimensional model that accurately represents a solid object.

Dear Family,

Your child will be learning to identify transformations on different figures. One special transformation that changes the size, but not the shape of a figure, is called a **dilation.** Typically, **enlargements** or **reductions** are types of dilations. Your child may use a **scale factor** to express how much an object is enlarged or reduced.

Dilate the figure by a scale factor of 3 and with _T_ as the center of dilation.

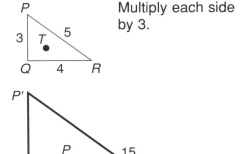

Multiply each side by 3.

Similar figures also use scale factors. **Similar figures** are figures that are the same shape, but not necessarily the same size. However, the angles of the figures do have to be congruent, and the ratios of corresponding sides must be equivalent. These ratios create the scale factor. Since the sides of similar figures are proportional, you can find an unknown dimension by using the properties of proportions.

Sam needs to pack a jewelry box that measures 4.5 inches wide and 6 inches long. If he finds a box that is similar with a length of 10 inches, how wide is the box?

$$\frac{4.5 \text{ in.}}{6 \text{ in.}} = \frac{x \text{ in.}}{10 \text{ in.}}$$ Set up a proportion.

$6 \bullet x = 4.5 \bullet 10$ Find the cross products.

$6x = 45$ Multiply.

$x = \frac{45}{6} = 7.5$ Solve for x.

The width of the box is 7.5 inches.

Holt Pre-Algebra

When an item is too large to view on paper, you need to make a scale drawing or scale model of the item. A **scale drawing** is an accurate two-dimensional representation of an object. A **scale model** is a three-dimensional representation of the actual object. Both the scale model and drawing are similar to the actual object.

A **scale** is used to show the ratio between the dimensions of the scale drawing or model and the actual object. Your child will learn to identify and use the scale to find the dimensions of a scale drawing, model, or actual object.

The length of an object on a scale drawing is 4 cm and its actual length is 400 m.

The scale is 1 cm: _____ m. What is the scale?

$$\frac{1 \text{ cm}}{x \text{ m}} = \frac{4 \text{ cm}}{400 \text{ m}}$$ Set up a proportion: $\frac{\text{scale length}}{\text{actual length}}$

$400 \cdot 1 = 4 \cdot x$ Find the cross products.

$x = 100$ Solve the proportion.

The scale is 1 cm:100 m.

A model of a 16-foot boat was made using the scale 3 in.:4 ft. What is the height of the model?

$$\frac{3 \text{ in.}}{4 \text{ ft}} = \frac{3 \text{ in.}}{48 \text{ in.}} = \frac{1 \text{ in.}}{16 \text{ in.}}$$ First find the scale factor.

Now that you have the scale factor, you can set up the proportion.

$$\frac{1}{16} = \frac{h \text{ in.}}{192 \text{ in.}}$$ Convert: 16 ft = 192 in.

$16h = 192$ Cross multiply.

$h = 12$ Solve for the height.

The height of the model is 12 in.

The material in this section has many real life applications. Have your child explain how to use a scale factor in relation to models and scale drawings. Practice having your child convert model dimensions to actual dimensions.

Sincerely,

Holt Pre-Algebra

Name _____ Date _____ Class _____

Tell whether each transformation is a dilation.

1.

2.
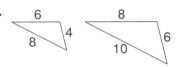

_____   _____

Use the properties of similar figures to answer each question.

3. A 5 in. long by 7 in. wide picture is going to be made into a similar poster with a length of 6 ft. How wide will the poster be?

_____

4. Is rectangle A similar to rectangle B or to rectangle C?

_____

5. Using a scale of $\frac{1}{2}$ cm:2 m, how long is an object that measures 3.5 cm long in a scale drawing?

_____

6. Using a 100x magnification microscope, a paramecium has a length of 2.8 mm. What is the actual length of the paramecium?

_____

Tell whether the scale reduces, enlarges, or preserves the actual object's size.

7. 3 in.:1 ft   8. 1 yd:36 in.   9. 5 m:1 m

_____   _____   _____

A cube with side lengths of 8 cm is built from small unit cubes. Compare the following values:

10. the side lengths of the two cubes

11. the surface areas of the two cubes

12. the volumes of the two cubes

_____   _____   _____

Answers: 1. yes **2.** no **3.** 8.4 ft **4.** rectangle C. **5.** 14 m **6.** 0.028 mm **7.** The scale is $\frac{1}{4}$ the size of the actual object. **8.** The scale is full size since the scale factor is 1. **9.** The scale enlarges the size of the object 5 times. **10.** The sides of the larger cube are 8 times longer than the smaller cube. **11.** The surface area of the larger cube is 64 times that of the smaller cube. **12.** The volume of the larger cube is 512 times that of the smaller cube.

55

Holt Pre-Algebra

Family Fun

Scrambled Fun

Why do Tommy Triangle's twin brothers love math class?

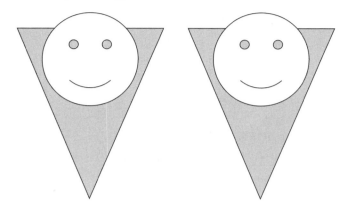

Directions

Unscramble each of the clue words.

Take the letters that appear in ⬜ boxes and unscramble them to get the answer to the riddle.

IOTAR

ROOPITRNPO

TERA

ROCSS RUDPOCT

OTALDIIN

SECLA RFTCAO

NERMEGLETNA

ERINDOTUC

CELSA LEMDO

CYPCIATA

B _ _ _ _ _ _   H _ _ _ _ _   _ _

_ _ _ _ _ _ _

Answer: ratio, proportion, rate, cross product, dilation, scale factor, enlargement, reduction, scale model, capacity; Because they are so similar.

Holt Pre-Algebra

Family Letter
Section A

What We Are Learning

Numbers and Percents

Vocabulary
These are the math words we are learning:

Percent A ratio that compares a number to 100. The symbol for percent is %.

Dear Family,

Your child will be learning about the relationships between decimals, fractions, and percents. To convert a fraction to a decimal, your child will divide the numerator by the denominator. To convert a decimal to a **percent,** your child will learn to multiply the decimal by 100 and add the percent sign.

Tom has a 5-gallon gas can that only has 2 gallons of gas in it. What percent of gas is in the gas can?

$\dfrac{\text{Gas in the can}}{\text{Capacity of the can}} = \dfrac{2}{5}$ Set up a ratio.

$\dfrac{2}{5} = 2 \div 5 = 0.40 = 40\%$ Find the percent.

Tom's gas can is 40% full.

Your child will be solving many different types of percent problems. One type of problem involves finding the percent of a number. Your child will learn two methods of trying to solve this type of problem as shown in the following example. Both methods are effective and acceptable.

What percent of 756 is 189?

Method 1: Set up an equation to find the percent.	Method 2: Set up a proportion to find the percent.
$p \cdot 756 = 189$ Set up an equation. $p = \dfrac{189}{756}$ Solve for p. $p = 0.25 \rightarrow 25\%$	*Think: What number is to 100 as 189 is to 756?* Set up the proportion. $\dfrac{x}{100} = \dfrac{189}{756}$ $756x = 18{,}900$ Solve for x. $x = 25\%$

As you can see, both methods provided the same results: 189 is 25% of 756.

Holt Pre-Algebra

The second type of percent problem is finding a number when the percent is known. In this case, your child will need to convert the percent to a decimal in order to multiply the two known values together.

What is 45% of 60?

$n = 45\% \cdot 60$	Set up an equation.
$n = 0.45 \cdot 60$	45% is equivalent to 0.45.
$n = 27$	Solve for n.

45% of 60 is 27.

The last type of percent problem your child will encounter in this section is finding the number that is multiplied by a percent to obtain a given number.

30 is 60% of what number?

Think: 60 is to 100 as 30 is to what number?

$\dfrac{60}{100} = \dfrac{30}{n}$	Set up the proportion.
$60n = 3000$	Find the cross products.
$n = 50$	Solve for n.

30 is 60% of 50.

Knowing the meaning of key words like "of" and "is" will help your child successfully set up the proportions or equations to find the missing value. Ask your child to explain how to solve each of the three types of percent problems.

Practice solving each type of problem on a regular basis to help build mastery with this concept.

Sincerely,

Holt Pre-Algebra

Family Letter

CHAPTER
8 *Numbers and Percents*

Find the missing ratio or percent equivalent for each letter on the number line.

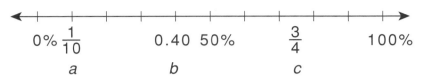

0% $\frac{1}{10}$ 0.40 50% $\frac{3}{4}$ 100%

    a b c

1. a **2.** b **3.** c

_____   _____   _____

Rewrite each value as indicated.

4. 58% as a decimal   **5.** $\frac{6}{25}$ as a percent   **6.** 0.784 as a fraction

_____   _____   _____

Find each percent.

7. 7.5 is what percent of 30?   **8.** What percent of 85 is 34?

_____   _____

9. What percent of 75 is 225?   **10.** 27 is what percent of 90?

_____   _____

11. About 635 students attend Golden State School. If 508 students ride
the bus to school, find the percent of students that ride the bus.

_____

Find each number.

12. 120 is $66\frac{2}{3}\%$ of what number?   **13.** 70% of what number is 168?

_____   _____

14. 96 is $12\frac{1}{2}\%$ of what number?   **15.** 42% of what number is 126?

_____   _____

16. In Mr. O'Riley's math classes, 32 students received A's and B's for
their third quarter grade. If this represents 64% of all of his students,
how many students does Mr. O'Riley teach?

_____

Answers: 1. 10% **2.** $\frac{2}{5}$ **3.** 75% **4.** 0.58 **5.** 24% **6.** $\frac{98}{125}$ **7.** 25% **8.** 40% **9.** 300% **10.** 30% **11.** 80% **12.** 180 **13.** 240 **14.** 768 **15.** 300 **16.** 50 students

Holt Pre-Algebra

CHAPTER 8 — Family Fun

Percent Puzzle

Directions

Each puzzle has 3 pieces.

- Cut out each puzzle piece and shuffle the pieces. Deal out three pieces to each player.

- To complete a puzzle, each piece must be equal to the other two pieces of the puzzle.

- If the three pieces do not make a match, you can discard one of your cards back to the deck and draw another card, or you can ask another player to give you the card you need to make your match. If you exchange a card with another player, you must give that player one of your cards.

- The first player to have the most completed puzzles wins the game.

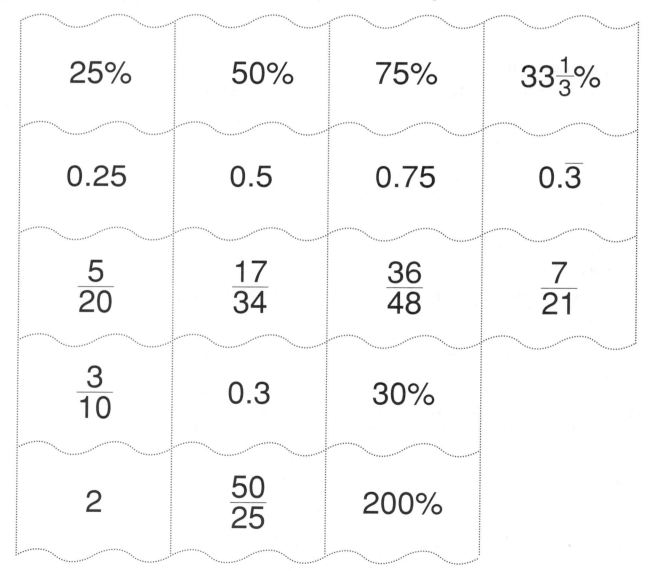

25%	50%	75%	$33\frac{1}{3}\%$
0.25	0.5	0.75	$0.\overline{3}$
$\frac{5}{20}$	$\frac{17}{34}$	$\frac{36}{48}$	$\frac{7}{21}$
$\frac{3}{10}$	0.3	30%	
2	$\frac{50}{25}$	200%	

Holt Pre-Algebra

What We Are Learning

Applying Percents

Vocabulary
These are the math words we are learning:

Commission A fee paid to a person who makes a sale.

Commission Rate The percent of the selling price a person earns for a sale.

Compatible Numbers Numbers that are close to the given numbers that make estimation or mental calculations easier.

Estimate A logical guess to the solution of a problem.

Interest The amount of money charged for borrowing or using money.

Percent of Change The percent by which a number increases or decreases.

Percent of Decrease A percent change describing a decrease in a quantity.

Percent of Increase The percent change describing an increase in a quantity.

Principal The initial amount of money borrowed or saved.

Rate of Interest The percent charged or earned on an amount of money.

Dear Family,

Your child will continue the study of percents by solving different application problems that involve percents. One everyday application is finding the **percent of change,** the rate of the amount of change to the original amount.

When finding the percent change, your child will either find the percent increase or the percent decrease.

Find the percent increase or decrease from 60 to 45, to the nearest percent.

First find the amount of change.

$60 - 45 = 15$ This is a percent decrease.

Think: What percent is 15 of 60?

$$\frac{\text{amount of decrease}}{\text{original amount}} \rightarrow \frac{15}{60} \qquad \text{Set up the ratio.}$$

$$\frac{15}{60} = \frac{1}{4} = 0.25 \qquad \text{Find the decimal form.}$$

$$= 25\% \qquad \text{Write as a percent.}$$

From 60 to 45 is a 25% decrease.

Your child will also learn to estimate with percents by using compatible numbers. **Compatible numbers** make it easier to estimate by providing numbers that go well together because they have common factors.

Estimate 37% of 25.

Instead of computing the exact answer of 37% • 25, estimate.

$$37\% = \frac{37}{100} \approx \frac{40}{100} \qquad \text{Use compatible numbers, 40 and 100.}$$

$$\approx \frac{2}{5} \qquad \text{Simplify.}$$

$$\frac{2}{5} \cdot 25 = 10 \qquad \text{Use mental math: } 2 \cdot 25 \div 5$$

37% of 25 is about 10.

Holt Pre-Algebra

Sales Tax   The tax on the sale of an item. It is a percent of the purchase price and is collected by the seller.

Simple Interest   A fixed percent of the principal.

Withholding Tax   A tax deducted from a person's earnings as an advance payment of income tax.

Another application of percent is calculating **commissions, sales tax,** and **withholding tax.** The method to calculate these amounts is very similar.

In some occupations, people are paid a percentage of a sale. This percentage amount is called a **commission rate.**

A real estate agent is paid a 4% commission on a house that sold for $125,000. How much was his commission?

Think: commission rate • sales = commission

$4\% \cdot 125{,}000 = c$	Write an equation.
$0.04 \cdot 125{,}000 = c$	Change the percent to a decimal.
$5000 = c$	Solve for c.

The agent earned a commission of $5000 on the sale.

Another real life application with percents involves **simple interest.** Your child will use the formula $I = P \cdot r \cdot t$ to find the simple interest on a specified amount of money.

Chas borrowed $3500 for 5 years at an annual simple interest rate of 5%. How much interest will he pay if he pays off the entire loan at the end of the fifth year?

$I = P \cdot r \cdot t$	Use the formula.
$I = 3500 \cdot 0.05 \cdot 5$	Substitute. Use 0.05 for 5%.
$I = 875$	Solve for I.

Chas will pay $875 in interest.

This section covers many applications of percents. Discuss with your child how percents are used in your family. If your child has a savings account, help calculate the interest that is being accrued each year. Being able to relate math to everyday situations will help your child see the connection between the concept and the application of mathematics.

Sincerely,

Holt Pre-Algebra

CHAPTER 8 **Family Letter**
Applying Percents

Find the percent of increase or decrease.

1. A coat was reduced from $70 to $45. Find the percent of decrease.

   _____

2. When Andre was 2 years old, he weighed 28 lb. At the age of 3, Andre weighed 34 lb. Find the percent of increase.

   _____

Estimate.

3. 16% of 40   **4.** 29% of 118   **5.** 96 out of 1020

_____   _____   _____

Solve.

6. Matthew works in a bicycle shop where he earns an 8% commission and no weekly salary. What will Matthew's weekly sales need to be to earn $450?

   _____

7. The sales tax in Sandra's town is 6.5%. If she buys 2 shirts for $23.99 each, and a pair of shoes for $41.99, how much sales tax does she owe?

   _____

8. Mark borrows $6000 from the bank at a simple interest rate of 7.5% for 5 years. How much will Mark pay back to the bank?

   _____

9. Marissa has a savings account with a balance of $2750. If the account earns an annual simple interest rate of 4.85%, how much will be in the account in 3 years?

   _____

10. Kevin earns $1066 every two weeks. Of that, $213.20 is withheld for taxes. What percent of Kevin's earnings is withheld each pay period?

   _____

11. A bank loaned a small business $55,000 at an annual simple interest rate. After 5 years, the business repaid the bank $60,345. What was the interest rate?

   _____

Answers: 1. 35.71% 2. 21.43% 3. 6 4. 36 5. $\frac{1}{10}$ or 10% 6. $5625 7. $5.85 8. $8250 9. $3150.13 10. 20%
11. 1.94%

CHAPTER 8

Family Fun
Money Matters

Materials
Number cube
Calculator

Directions

- Cut out the cards and place in a stack face down.

- Divide into teams. Each team is given a beginning balance of $500.

- Roll a number cube. If your team rolls an even number, you will earn a commission. If it is an odd number you will be taxed.

- Draw a card to determine your commission rate or tax rate. Determine your commission or tax by multiplying the rate times your current balance at the end of each round.

- Add your commission to the running balance. Subtract the tax from the running balance.

- The game lasts ten rounds. After the final round, the team with the highest ending balance is the winner.

5%	10%	15%	20%	25%
30%	5%	10%	15%	20%

Round	Rate	Balance $500	Round	Rate	Balance $500
1			6		
2			7		
3			8		
4			9		
5			10		

Holt Pre-Algebra

Family Letter
Section A

What We Are Learning

Experimental Probability

Vocabulary
These are the math words we are learning:

Certain An event having a probability of 1.

Event A set of one or more outcomes of an experiment.

Experiment An activity based on chance.

Experimental Probability The ratio of the number of times an event occurs to the total number of trials.

Impossible An event having a probability of 0.

Outcome A possible result of an experiment.

Probability A number from 0 to 1 that tells how likely an event is to happen.

Random Numbers A set of numbers in which there is no pattern that can be used to predict the next number.

Sample Space The set of all possible outcomes of an experiment.

Simulation A model of a real situation.

Trial The act of trying, testing, or putting to the proof.

Dear Family,

In this section your child will be learning the necessary terminology for understanding and applying the concepts of **probability.**

Here is how your child will learn how to find the probabilities of **outcomes** in a **sample space.**

A number cube has a digit from 1 to 6 written on each of the faces. So, the sample space is {1, 2, 3, 4, 5, 6}.

A. What is the probability of rolling a 4?
There is only one 4 and there are a total of 6 digits in the sample space.

$P(\text{rolling a 4}) = \frac{1}{6}$.

B. What is the probability of not rolling a 4?
The probabilities must add up to 1.

$P(\text{not rolling a 4}) = 1 - P(\text{rolling a 4}) = 1 - \frac{1}{6} = \frac{5}{6}$.

Your child will also learn how to find probabilities of **events** by adding the probabilities of all the outcomes included in the event.

A bank teller's cash drawer contains 4 types of bills; $1, $5, $10, and $20. The table gives the probability of randomly selecting each type of bill from the drawer.

Bill	$1	$5	$10	$20
Probability	0.36	0.30	0.24	0.10

A. What is the probability of selecting a $1 or a $5?
$P(\$1 \text{ or } \$5) = P(\$1) + P(\$5) = 0.36 + 0.30 = 0.66$

B. What is the probability of not selecting a $1?
Since the probabilities must add up to 1, you can subtract.

$P(\text{not selecting } \$1) = 1 - P(\text{selecting } \$1) = 1 - 0.36 = 0.64$

Holt Pre-Algebra

Your child will also be introduced to **experimental probability**.

A large aquarium contains goldfish, black fish, and red fish. Julie recently has seen 5 goldfish, 7 black fish and 3 red fish. What is the probability that the next fish she sees will be a goldfish?

Outcomes	Gold	Black	Red
Observations	5	7	3

$$\frac{\text{number of goldfish}}{\text{total number of fish}} = \frac{5}{5 + 7 + 3} = \frac{5}{15} = \frac{1}{3}$$

The probability that Julie sees a goldfish next is $\frac{1}{3}$ or 33%.

Your child will also learn to use a set of **random numbers** to **simulate** a situation and find probabilities.

The weatherman, Joe Blizzard, is famous because he is able to correctly predict the coming of a storm 92% of the time. In July, there were 6 storms. Estimate the probability that Joe predicted at least 5 correctly. Use 5 trials for the simulation.

The table below is a set of random digits. Use digits from the table, grouped in pairs, to simulate the situation.

94182 73266 67899 38783 94228 23426 76679 41256
39917 16373 98733 18594 22545 61378 33563 65161

Since Joe is accurate 92% of the time, the numbers 01–92 will represent a correct prediction. The numbers 93–00 will represent an incorrect prediction. Since there were 6 storms, you will need 6 pairs of numbers in each of the 5 trials. Each trial is listed below with a correct prediction in gray.

94	18	27	32	66	67	5 correct predictions
89	93	87	83	94	22	4 correct predictions
82	34	26	76	67	94	5 correct predictions
12	56	39	91	71	63	6 correct predictions
73	98	73	31	85	94	4 correct predictions

Out of 5 trials, 3 represented 5 or more correct predictions. Based on the simulation, the probability of Joe predicting at least 5 correctly is $\frac{3}{5}$ or 60%.

Help your child understand probability by relating it to everyday events.

Sincerely,

Holt Pre-Algebra

CHAPTER 9 **Family Letter**

Experimental Probability

Give the probability of each outcome or event.

1. rolling a 2 on a number cube   _____

2. rolling an odd number on a number cube   _____

An experiment consists of reaching into a bag, pulling out a handful of mixed coins, and counting the number of pennies. The table gives the probability of each outcome.

Outcome	0	1	2	3
Probability	0.14	0.36	0.41	0.09

3. What is the probability of pulling out exactly 1 penny?   _____

4. What is the probability of pulling out at least 1 penny?   _____

5. What is the probability of not pulling out a penny?   _____

Maria is counting the number of red cars, black cars, and white cars that pass her house. Her count is recorded in the table.

Outcome	Red	Black	White
Number of cars	32	8	24

6. Estimate the probability that the next car she sees will be white.   _____

7. Estimate the probability that the next car will be black or white.   _____

8. Estimate the probability that the next car will not be white.   _____

Use the table of random numbers to simulate the situation. Use at least 10 trials.

49064 63415 12830 11776 66783 31260 14965 42974 24935 65178 17252 33498
52737 11152 02985 12526 15735 21997 56472 34982 26548 15145 59731 55684
24976 53846 22498 56731 56789 02457 56821 79056 21934 79165 50244 65178

9. Surveys show that 72% of adults do not like horror movies. Estimate the probability that at least the next 4 out of 5 adults that walk in a movie theater will not like horror movies.

Answers: 1. $\frac{1}{6}$ 2. $\frac{1}{2}$ 3. 0.36 4. 0.86 5. 0.14 6. $\frac{3}{8}$ 7. $\frac{1}{2}$ 8. $\frac{5}{8}$ 9. $\frac{6}{10}$ or 60%

Holt Pre-Algebra

CHAPTER 9	**Family Fun**

Probability Word Search

Directions

- Fill in the blanks below with the appropriate vocabulary words.

- Locate each of the words in the word search puzzle and circle it.

- The player who finds all eleven words first wins the game.

1. An _____ is a set of one or more outcomes.

2. The _____ of an event is a number from 0 to 1 that tells how likely the event is to happen.

3. An _____ is an activity in which results are observed.

4. A result of an experiment is an _____.

5. A probability of 1 means that an event is _____.

6. A probability of 0 means that an event is _____.

7. The _____ is the set of all possible outcomes to an experiment.

8. Each observation of a test is called a _____.

9. _____ probability is the ratio of the number of times an event occurs to the number of trials.

10. A _____ is a model of a real situation.

11. _____ numbers are a set of numbers in which there is no pattern.

```
E  C  A  P  S  E  L  P  M  A  S  X  W  O  E
S  X  K  I  E  E  F  H  L  E  A  T  B  U  J
I  M  P  O  S  S  I  B  L  E  Z  I  N  T  L
M  H  A  E  H  L  O  E  J  T  M  I  G  C  B
U  S  B  M  R  H  U  P  W  S  A  F  H  O  M
L  H  Q  D  A  I  M  G  L  T  Y  H  X  M  U
A  L  Z  T  N  E  M  I  R  E  P  X  E  E  L
T  B  J  K  D  O  G  E  V  E  N  T  J  N  A
I  L  M  H  O  E  C  A  N  E  P  R  D  L  L
O  P  S  I  M  E  T  K  X  T  N  I  D  H  M
N  T  E  L  W  O  D  A  N  T  A  A  K  Q  O
A  E  H  M  P  R  O  B  A  B  I  L  I  T  Y
```

Answers: 1. event 2. probability 3. experiment 4. outcome 5. certain 6. impossible 7. sample space 8. trial 9. Experimental 10. simulation 11. Random

Holt Pre-Algebra

What We Are Learning

Theoretical Probability and Counting

Vocabulary

These are the math words we are learning:

Combination An arrangement of items or events in which order does not matter.

Dependent Events Events for which the outcome of the first event affects the outcome of the second event.

Equally Likely Outcomes that have the same probability.

Factorial The product of all whole numbers, except zero, less than or equal to a number.

Fair When all outcomes of an experiment are equally likely.

Fundamental Counting Principle If there are m ways to choose a first item and n ways to choose a second item after the first item has been chosen, then there are $m \cdot n$ ways to choose both items.

Independent Events The occurrence of one event has no effect on the probability that a second event will occur.

Dear Family,

In this section your child will learn how to apply the concepts of probability. This is how your child will apply the concepts of **theoretical probability**.

An experiment consist of rolling a die and flipping a coin. There are 12 possible outcomes: H1, H2, H3, H4, H5, H6, T1, T2, T3, T4, T5, and T6. What is the probability of getting a heads and rolling an odd number?

There are 3 outcomes in the event of getting a heads and rolling an odd number: H1, H3, and H5.

$P(\text{heads and odd}) = \frac{3}{12} = \frac{1}{4}$.

Your child will also be introduced to the **Fundamental Counting Principle** in order to find the total number of ways that two or more separate events can happen.

A game consists of rolling a die and spinning a spinner that is divided into equal quarters of different colors. How many different outcomes in one turn are possible?

Number of outcomes when rolling a die: 6
Number of outcomes when spinning the spinner: 4
$6 \cdot 4 = 24$. There are 24 possible outcomes in any one turn.

Your child will learn the difference between **permutations** and **combinations** and how to evaluate each.

You are required to read 3 books this summer. You can choose from a list of 10 books. How many different ways can you choose the 3 books?

Since order does not matter, this is a combination.

$$_{10}C_3 = \frac{10!}{3!(10-3)!} = \frac{10 \cdot 9 \cdot 8 \cdot 7 \cdot 6 \cdot 5 \cdot 4 \cdot 3 \cdot 2 \cdot 1}{3 \cdot 2 \cdot 1 \cdot 7 \cdot 6 \cdot 5 \cdot 4 \cdot 3 \cdot 2 \cdot 1} = 120$$

If you must read one in June, one in July, and one in August, how many different ways can you do this?

Since order does matter, this is a permutation.

$$_{10}P_3 = \frac{10!}{(10-3)!} = \frac{10 \cdot 9 \cdot 8 \cdot 7 \cdot 6 \cdot 5 \cdot 4 \cdot 3 \cdot 2 \cdot 1}{7 \cdot 6 \cdot 5 \cdot 4 \cdot 3 \cdot 2 \cdot 1} = 720$$

Holt Pre-Algebra

Family Letter

Section B, continued

Mutually Exclusive
Events that cannot occur in the same trial of an experiment.

Odds Against The ratio of unfavorable outcomes to favorable outcomes.

Odds in Favor The ratio of favorable outcomes to unfavorable outcomes.

Permutation An arrangement of events in a certain order.

Theoretical Probability
The ratio of the number of favorable outcomes to the number of possible outcomes.

Tree Diagram A branching diagram that shows all the possible outcomes of an event.

To find the probability that two **independent events** occur, your child will multiply the probability that the first event happens by the probability that the second event happens.

Joey is holding two hats. Both contain 26 pieces of paper on which each letter of the alphabet is written. If you choose a letter from each hat, what is the probability that you choose a vowel from each?

$$P(\text{vowel from 1}^{st}\text{ hat}) = \frac{5 \text{ vowels}}{26 \text{ letters}}$$

$$P(\text{vowel from 2}^{nd}\text{ hat}) = \frac{5 \text{ vowels}}{26 \text{ letters}}$$

$$P(\text{vowel and vowel}) = \frac{5}{26} \cdot \frac{5}{26} = \frac{25}{676}$$

To find the probability that two **dependent events** occur, your child will multiply the probability that the first event happens by the probability that the second event happens. Students must keep in mind that the first event will affect the probability of the second event.

Joey is now holding one hat containing 26 pieces of paper on which each letter of the alphabet is written. If you choose 2 letters from the hat, what is the probability that both are vowels?

$$P(\text{vowel from 1}^{st}\text{ choice}) = \frac{5 \text{ vowels}}{26 \text{ letters}}$$

$$P(\text{vowel from 2}^{nd}\text{ choice}) = \frac{4 \text{ vowels left}}{25 \text{ letters left}}$$

$$P(\text{vowel and vowel}) = \frac{5}{26} \cdot \frac{4}{25} = \frac{2}{65}$$

Your child will learn how to calculate both the odds in favor of an event and the odds against an event happening.

odds in favor = number of favorable outcomes:number of unfavorable outcomes

odds against = number of unfavorable outcomes:number of favorable outcomes

Try to relate everyday probability situations to your child to help reinforce the ideas and concepts in this lesson.

Sincerely,

Holt Pre-Algebra

CHAPTER 9 — Family Letter
Theoretical Probability and Counting

An experiment consists of rolling a fair die and flipping a coin. Find each probability.

1. rolling a 5 _____

2. rolling a number less than 5 and getting a heads _____

Find the number of possible outcomes.

3. A pizza shop offers thick crusts, thin crusts, and stuffed crusts. The choices of toppings are pepperoni, cheese, hamburger, peppers, sausage, onions and mushrooms. How many different one-topping pizzas can you order? _____

4. If a code consists of 3 letters, followed by 2 numbers, how many different codes can be made? _____

Evaluate each expression.

5. $_6C_2$

_____

6. $_8P_3$

_____

A bag contains 7 red marbles, 2 blue marbles, and 1 green marble. Find each probability.

7. Find the probability of drawing a red marble, replacing it, and drawing another red marble. _____

8. Find the probability of drawing two red marbles if you do not replace the first marble. _____

Of the first 1000 people to arrive at a concert, 25 win door prizes.

9. Estimate the odds in favor of winning a door prize. _____

10. Estimate the odds against winning a door prize. _____

Answers: 1. $\frac{1}{6}$ 2. $\frac{4}{12} = \frac{1}{3}$ 3. 3 • 7 = 21 4. 26 • 26 • 10 • 10 = 1,757,600 5. 15 6. 336 7. $\frac{7}{10} • \frac{7}{10} = \frac{49}{100}$ 8. $\frac{7}{10} • \frac{6}{9} = \frac{42}{90} = \frac{7}{15}$ 9. 25:975 10. 975:25

Holt Pre-Algebra

Family Fun

Quick Combinations

Directions

• Cut out the cards below and place them face up on a table.

• One player at a time will try and make 10 possible 3-number combinations out of the numbers below. Another player should record the combinations in the table to the right as well as the time.

• The player should keep making combinations until they find all 10. Remember, order does not matter in combinations, so 123 is the same as 321, etc.

• The player with the fastest time wins the game.

	Player 1	Player 2	Player 3
1			
2			
3			
4			
5			
6			
7			
8			
9			
10			
Time			

1   **2**

3   **4**   **5**

Answer: The ten different combinations in any order are: 123, 124, 125, 134, 135, 145, 234, 235, 245, 345

Holt Pre-Algebra

What We Are Learning

Solving Linear Equations

Dear Family,

In this section your child will be learning to solve equations by isolating the variable and finding its value.

The easiest equations to solve are those that have only one variable term. Your child will learn to solve these types of equations by performing inverse (opposite) operations on each of the numbers that are on the same side as the variable.

Solve. $-4x + 8 = 16$

$$-4x + 8 = 16$$
$$\underline{\quad -8 \quad -8 \quad}$$
$$-4x = 8$$

Get the x by itself.
First subtract 8 from both sides.

$$\frac{-4x}{-4} = \frac{8}{-4}$$

Next, divide both sides by -4.

$$x = -2$$

Check your answer by substituting -2 for x.

$$-4(-2) + 8 = 8 + 8 = 16 \ ✔$$

If an equation has a variable in the numerator or denominator of a fraction, your child will need to "clear" the fraction by multiplying each term by the denominator of the fraction. This will make the problem less complicated to solve.

Solve. $\dfrac{c + 5}{3} = 7$

$$3 \cdot \frac{c + 5}{3} = 7 \cdot 3$$

Multiply both sides of the equation by 3.

$$c + 5 = 21$$

$$c + 5 = 21$$
$$\underline{\quad -5 \quad -5 \quad}$$

Subtract 5 from both sides.

$$c = 16$$

Holt Pre-Algebra

Your child will also learn to solve equations where there is more than one variable term on the same side of an equation. The first step is to combine like terms so that there is only one variable term. Then solve for the variable using inverse operations.

Solve. $3k + 5 - k = -9$

$3k + 5 - k = -9$	Combine "like terms." $3k - k = 2k$
$2k + 5 = -9$	
$2k + 5 = -9$	Now get the k alone.
$\underline{ -5 \quad -5}$	
$2k = -14$	Subtract first.
$\dfrac{2k}{2} = \dfrac{-14}{2}$	Divide next.
$k = -7$	

Your child will also learn to solve equations with variables on both sides of the equal sign. The first goal is to get all of the variables on one side by using inverse operations. Then, the steps are the same as in the previous problems.

Solve. $6y - 5 = 3 + 4y$

$6y - 5 = 3 + 4y$	Get all y's on one side by
$\underline{-4y \qquad\qquad -4y}$	subtracting $4y$ from both sides.
$2y - 5 = 3$	
$2y - 5 = 3$	
$\underline{ +5 \quad +5}$	Add 5 to both sides.
$2y = 8$	
$\dfrac{2y}{2} = \dfrac{8}{2}$	Divide both sides by 2.
$y = 4$	

It is important that your child learn to solve equations. He or she will use this skill in every math course they take from this point forward. Reinforce these concepts by practicing regularly.

Sincerely,

Holt Pre-Algebra

Name _____ Date _____ Class _____

Family Letter
Solving Linear Equations

Solve. Check each answer.

1. $4x - 7 = 17$

2. $\dfrac{h}{6} - 11 = -14$

3. $\dfrac{y + 5}{6} = 4$

_____

_____

_____

4. $19 = 7 - 3x$

5. $\dfrac{3 - 6k}{3} = -9$

6. $5 + \dfrac{m}{3} = 1$

_____

_____

_____

7. $9n - 5 + 3n = -5$

8. $8p - 9 + 6p - 7 = -2$

9. $\dfrac{d}{5} - \dfrac{1}{5} = \dfrac{14}{5}$

_____

_____

_____

10. $5(2w + 1) + w = 49$

11. $15x + 9 - 2x = -30$

12. $\dfrac{3}{7} + \dfrac{2p}{7} = \dfrac{15}{7}$

_____

_____

_____

13. $13h = 45 - 2h$

14. $8 + 6k = -10 - 24$

15. $4(x + 5) = 15 - x$

_____

_____

_____

16. $4(n - 5) = 9n$

17. $-21 + 3b = -11b + 7$

18. $2(5 + 3x) = 4(2x)$

_____

_____

_____

Answers: 1. $x = 6$ 2. $h = -18$ 3. $y = 19$ 4. $x = -4$ 5. $k = 5$ 6. $m = -12$ 7. $n = 0$ 8. $p = 1$ 9. $d = 15$
10. $w = 4$ 11. $x = -3$ 12. $p = 6$ 13. $h = 3$ 14. $k = -7$ 15. $x = -1$ 16. $n = -4$ 17. $b = 2$ 18. $x = 5$

Holt Pre-Algebra

CHAPTER	**Family Fun**
10	*Four in a Row Equations*

Directions

• Each player should create a 4–by–4 grid as shown.

• Have each player write the equations below in any order in each one of the squares in the grid.

$x = 0$ $x = 1$ $x = 2$ $x = 3$ $x = 4$

$x = 5$ $x = 6$ $x = 7$ $x = 8$

$x = -1$ $x = -2$ $x = -3$ $x = -4$

$x = -5$ $x = -6$ $x = -7$

• Each player will work out problems 1 to 16 in the order in which they appear. If the player finds the solution on his or her grid, the player will color in the corresponding square.

• The first player to color in four squares in a row, column, or diagonal wins the game.

1. $-5x + 1 = -15 + 1$

2. $3 = 5 - 2x$

3. $3x = 4x - 8$

4. $11(x + 4) = 44$

5. $-23 = 4x + 3x + 5$

6. $7x + 1 - 3x = 9$

7. $-15x + 1 = -13x + 15$

8. $3(x - 4) = 5x$

9. $\frac{5x}{7} + \frac{12}{7} = \frac{2}{7}$

10. $\frac{x - 8}{3} = -3$

11. $9(2x) = 4x + 21 + 3x + 23$

12. $2x + 1 = 4x + 7$

13. $\frac{2x + 60}{4} = 3x$

14. $\frac{3}{8} + \frac{x}{8} = \frac{10}{8}$

15. $5 - (x + 2) = -2x - 2$

16. $2x - 5 = x$

Answers: 1. $x = 3$ **2.** $x = 1$ **3.** $x = 8$ **4.** $x = 0$ **5.** $x = -4$ **6.** $x = 2$ **7.** $x = -7$ **8.** $x = -6$ **9.** $x = -2$ **10.** $x = -1$ **11.** $x = 4$ **12.** $x = -3$ **13.** $x = 6$ **14.** $x = 7$ **15.** $x = -5$ **16.** $x = 5$

Holt Pre-Algebra

Family Letter
Section B

What We Are Learning

Solving Equations and Inequalities

Vocabulary
These are the math words we are learning:

Solution of a System of Equations A set of values making all equations in a system true.

System of Equations A system of two or more equations that contain two or more variables.

Dear Family,

In this section your child will learn to solve inequalities by isolating the variable and finding its value. He or she will then learn to solve for a variable in equations with other variables. Finally, your child will learn to find the solution to a system of equations.

Your child will learn to find solution sets of inequalities using the same methods they used to find solutions to equations, by isolating the variable.

As with equations, when isolating the variable they should follow the order of operations in reverse (subtraction and addition first, division and multiplication next).

Solve. Then graph the solution set on a number line.

$\frac{x}{4} + 7 \le 12$

$\underline{-7 -7 }$ First, subtract 7 from both sides.

$\frac{x}{4} \le 5$ Next, multiply both sides by 4.

$4 \cdot \frac{x}{4} \le 5 \cdot 4$

$x \le 20$

If a step in solving the inequality involves multiplying or dividing by a negative number, help your child to remember to switch the direction of the inequality symbol.

Solve. Then graph the solution set on a number line.

$11 - 6x < 23$

$\underline{-11 -11 }$ First, subtract 11 from both sides.

$-6x < 12$ Next, divide both sides by -6 and

$\frac{-6x}{-6} > \frac{12}{-6}$ reverse the inequality symbol.

$x > -2$

Holt Pre-Algebra

Family Letter
Section B, continued

In both math and science, there are times when a common equation is easier to use if a certain variable is alone on one side of the equal sign.

Solve $d = r \cdot t$ for t.

$d = r \cdot t$

$\dfrac{d}{r} = \dfrac{r \cdot t}{r}$ Divide both sides by r.

$\dfrac{d}{r} = t$ OR $t = \dfrac{d}{r}$ The t is now alone.

Your child will be introduced to **systems of equations** and learn how to find the values that make a system true. Students will learn to find the **solution to a system of equations** by first getting the same variable alone in both equations. Then students will set both of the new equations equal to each other and solve for the value of the other variable. Finally, students will substitute the value they found into either equation to find the value of the first variable.

Solve the system of equations. $x + 3y = 5$
$x + 2y = 3$

$\begin{aligned} x + 3y &= 5 \\ -3y \quad &-3y \\ \hline x &= 5 - 3y \end{aligned}$ $\begin{aligned} x + 2y &= 3 \\ -2y \quad &-2y \\ \hline x &= 3 - 2y \end{aligned}$ Get the x by itself in both equations.

$5 - 3y = 3 - 2y$ Set $x = x$.

$\begin{aligned} 5 - 3y &= 3 - 2y \\ +3y \quad &+3y \\ \hline 5 &= 3 + y \end{aligned}$ Simplify and solve for y.

$\begin{aligned} 5 &= 3 + y \\ -3 \quad &-3 \\ \hline 2 &= y \end{aligned}$ Subtract 3 from both sides

$x + 3(2) = 5$ Substitute 2 into either equation for y to find the value of x.

$x + 6 = 5$

$x = -1$ Solution: $(x, y) \rightarrow (-1, 2)$

Reinforce with your child the concepts introduced in this section by practicing the steps learned.

Sincerely,

Holt Pre-Algebra

Name _____ Date _____ Class _____

CHAPTER 10 **Family Letter**
Solving Equations and Inequalities

Solve and Graph.

1. $2x - 9 < 15$

2. $4 - 5b > 19$

3. $7p + 9 - 4p \geq 15$

4. $2(k + 1) < 6$

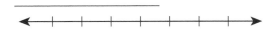

5. $3 > 4 - x$

6. $4 + 6y \leq -1 + 11y$

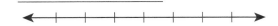

Solve each equation for the indicated variable.

7. Solve $x + y = 5$ for x. _____

8. Solve $d = rt$ for t. _____

9. Solve $V = \pi r^2 h$ for h. _____

10. Solve $P = 2\ell + 2w$ for ℓ. _____

11. Solve $y = 3x + 5$ for x. _____

Solve each system of equations.

12. $y = 3x + 9$
$y = -7x + 19$

13. $3x + 9y = -9$
$x - 2y = 7$

Answers: 1. $x < 12$ 2. $b < -3$ 3. $p \geq 2$ 4. $k < 2$ 5. $x > 1$ 6. $y \geq 1$ 7. $x = 5 - y$ 8. $t = \frac{d}{r}$ 9. $h = \frac{V}{\pi r^2}$ 10. $\ell = \frac{P - 2w}{2}$ 11. $x = \frac{y - 5}{3}$ 12. (1, 12) 13. (3, −2)

Holt Pre-Algebra

CHAPTER **10**

Family Fun
Crossword Inequalities

Directions

- Each player should solve the problems to the right and circle their answers.

- Using the across and down clues at the bottom of the page, each player should match each clue with one of their solutions and write the inequality or equation they solved to get that solution in the corresponding crossword boxes.

- Each digit, letter, and symbol should be placed in a separate box. Example: $24 + 3x \geq 21$ would need 8 boxes; one each for the 2, 4, +, 3, x, $\geq$, 2, and 1.

- A negative number only requires one box. Example: $-6 + x \geq 5$ would need 5 boxes; one each for the -6, +, x, $\geq$, and 5.

- In problems with directions, do not write the directions in the puzzle. Example: "Solve for x."

- When writing a system, write both equations next to each other, in the order given, without a space.

- The player who completes the crossword puzzle the fastest wins the game.

Problems
$14x + 2 > 30$
Solve for x. $z = y + x$
$-6 + x \geq 5$
Solve for x. $z = xy$
Solve the system. $x + y = 4$ $x - y = 6$
$23 + 6x > -3x + 5$
$24 + 3x < 21$
$3x \geq -6$
Solve the system. $y = x - 6$ $y = 2x + 3$
$2x - 3 < 7$
$7x - 1 \geq 6$

Across

1. $x = z - y$

2. $x > 2$

3. $x \geq -2$

4. $x < 5$

5. $x = -9, y = -15$

6. $x < -1$

Down

1. $x = \dfrac{z}{y}$

2. $x = 5, y = -1$

3. $x \geq 1$

4. $x \geq 11$

5. $x > -2$

What We Are Learning

Linear Equations

Vocabulary
These are the math words we are learning:

Linear Equation An equation whose solutions fall on a straight line on the coordinate plane.

Point-Slope Form The equation of a line in the form of $y - y_1 = m(x - x_1)$, where m is the slope and (x_1, y_1) is a specific point on the line.

Slope-Intercept Form A linear equation written in the form $y = mx + b$, where m is the slope and b is the y-intercept.

x-intercept The x-coordinate of the point where the line crosses the x-axis.

y-intercept The y-coordinate of the point where the line crosses the y-axis.

Dear Family,

Your child will be learning how to graph **linear equations** using a table, slope, and **x-** and **y-intercepts**. Students will also learn to write linear equations in **slope-intercept form** and in **point-slope form**.

To determine if an equation is linear, graph the equation using a table of values and then see if all the solutions fall on a straight line on the coordinate plane.

Graph the equation $y = -x + 5$ and tell if it is linear.

x	$-x + 5$	y	(x, y)
-2	$-(-2) + 5$	7	$(-2, 7)$
-1	$-(-1) + 5$	6	$(-1, 6)$
0	$-(0) + 5$	5	$(0, 5)$
1	$-(1) + 5$	4	$(1, 4)$
2	$-(2) + 5$	3	$(2, 3)$
3	$-(3) + 5$	2	$(3, 2)$

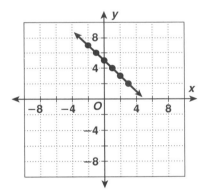

The equation $y = -x + 5$ is linear since its graph is a straight line.

Another linear equation concept that will be introduced to your child is slope of a line. Your child will use the formula

$m = \dfrac{y_2 - y_1}{x_2 - x_1}$ to find the slope of the line.

Find the slope of the line that passes through $(4, -1)$ and $(6, 2)$.

Let (x_1, y_1) be $(4, -1)$ and (x_2, y_2) be $(6, 2)$.

$m = \dfrac{y_2 - y_1}{x_2 - x_1} = \dfrac{2 - (-1)}{6 - 4} = \dfrac{3}{2}$

The slope of the line is $\dfrac{3}{2}$.

Holt Pre-Algebra

Shown below are the various types of slopes of a line.

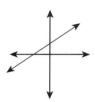 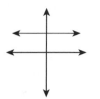

| Positive slope | Negative slope | Zero slope | Undefined slope |

Your child will use the slope-intercept form of an equation, $y = mx + b$, to determine the slope and y-intercept.

$$y = mx + b$$

slope y-intercept

Write the equation in slope-intercept form and then find the slope and y-intercept.

$-3x + 4y = 12$

$\underline{+3x \qquad\qquad +3x}$ Add $3x$ to both sides.

     $4y = 12 + 3x$

Rewrite to match slope-intercept form.

$4y = 12 + 3x$

$\dfrac{4y}{4} = \dfrac{12}{4} + \dfrac{3x}{4}$ Divide each side by 4.

$y = \dfrac{3}{4}x + 3$ The equation is in slope-intercept form.

$m = \dfrac{3}{4}, b = 3$

The slope of the line $-3x + 4y = 12$ is $\dfrac{3}{4}$ and the y-intercept is 3.

It is important that your child grasp the concepts of associated with linear equations. Make sure your child spends time on the homework assignments.

Sincerely,

Holt Pre-Algebra

CHAPTER **Family Letter**

11 *Linear Equations*

Graph each equation and tell whether it is linear.

1. $y = x + 3$ _____

2. $y = -3x$ _____

3. $y = x^2$ _____

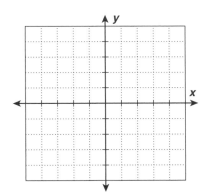
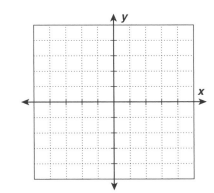
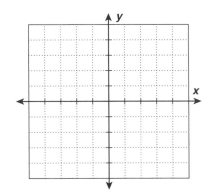

Use the graph of each line to determine its slope.

4. _____

5. _____

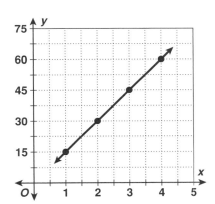
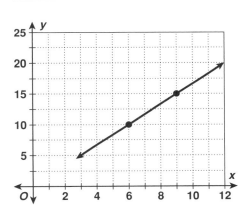

Write each equation in slope-intercept form and then find the slope and *y*-intercept.

6. $3x = 9y$ **7.** $8 = 2x + 8y$ **8.** $4x - 6y = 24$ **9.** $2 = y$

_____   _____   _____   _____

Write the point-slope form of the equation with the given slope that passes through the indicated point.

10. the line with a slope 3 passing through (2, 7)

11. the line with a slope 0 passing through (5, −1)

_____   _____

Answers: 1. linear **2.** linear **3.** nonlinear **4.** 15 **5.** $\frac{5}{3}$ **6.** $y = \frac{1}{3}x$; $m = \frac{1}{3}$; (0, 0) **7.** $y = 1 - \frac{1}{4}x$; $m = -\frac{1}{4}$; (0, 1) **8.** $y = \frac{2}{3}x - 4$; $m = \frac{2}{3}$; (0, −4) **9.** $y = 2$; $m = 0$; (0, 2) **10.** $y - 7 = 3(x - 2)$ **11.** $y + 1 = 0(x - 5)$

Holt Pre-Algebra

Family Fun
Hidden Phrases

What do Polly's friends say when she leaves the room?

Directions
- Determine the slope for each problem using the slope formula, slope-intercept form, or point-slope form.
- Find that number down below and match it with a letter.
- Put that letter in the correct place to solve the riddle.

_____ **1.** $(2, 3)$; $(1, 6) \rightarrow$ A

_____ **2.** $y = -2x + 5 \rightarrow$ P

_____ **3.** $(0, 0)$; $(4, -2) \rightarrow$ O

_____ **4.** $y - 1 = 3(x + 1) \rightarrow$ N

_____ **5.** $(-2, 5)$; $(-1, 1) \rightarrow$ M

_____ **6.** $(0, -6)$; $(5, 2) \rightarrow$ Y

_____ **7.** $y = 6 + x \rightarrow$ L

_____ **8.** $(-2, -2)$; $(7, -2) \rightarrow$ G

_____ **9.** $y + 4 = -\frac{1}{2}(x - 3) \rightarrow$ O

___ ___ ___ ___ ___ ___ ___

$$-2 \quad -\frac{1}{2} \quad 1 \quad \frac{8}{5} \quad 0 \quad -\frac{1}{2} \quad 3$$

Answer: 1. -3 **2.** -2 **3.** $-\frac{1}{2}$ **4.** 3 **5.** -4 **6.** $\frac{8}{5}$ **7.** 1 **8.** 0 **9.** $-\frac{1}{2}$ word: polygon

Holt Pre-Algebra

Family Letter

What We Are Learning

Linear Relations

Vocabulary
These are the math words we are learning:

Boundary Line The points where the two sides of a two-variable inequality are equal.

Constant of Proportionality The ratio by which two variable quantities are related.

Direct Variation A relationship between two variables such that the data increase or decrease together at a constant rate.

Linear Inequality When the equality symbol is replaced in a linear equation by an inequality symbol.

Dear Family,

Your child will be learning how to use **direct variation** to relate two variable quantities, graph inequalities in two variables, and reference lines of best fit.

You can determine if two variable quantities have a direct relationship if the increase (or decrease) in one variable increases (or decreases) the other variable.

Determine whether the data set shows direct variation.

Lincoln High School Baseball and Softball Games

Softball Hits Per Game	8	12	11	15	9
Baseball Hits Per Game	5	7	8	13	12

Compare the ratios to see if a direct variation occurs.

$$\frac{8}{5} \overset{?}{\underset{}{\times}} \frac{9}{12} \quad \boxed{\begin{array}{c} 45 \\ \hline 96 \end{array}}$$

This ratio is not constant because $45 \neq 96$. The relationship of the data is not a direct variation.

Your child will also be learning how to find the constant of variation in a direct variation relationship and how to write an equation of direct variation.

Find the equation of direct variation, given that y varies directly with x.
x is -4 when y is 14

$y = kx$	y varies directly with x.
$14 = k \cdot -4$	Substitute for x and y.
$-\dfrac{14}{4} = k$	Solve for k.
$-\dfrac{7}{2} = k$	Simplify.
$y = -\dfrac{7}{2}x$	Substitute $-\dfrac{7}{2}$ for k in the original equation.

Holt Pre-Algebra

In this section your child will be introduced to graphing inequalities in two variables. Students will learn to rewrite inequalities in slope-intercept form in order to graph linear inequalities, as shown below.

Graph the inequality.

$x + y > 5$

First, write the equation in slope-intercept form.

$$\begin{array}{ll} x + y > 5 & \\ \underline{-x \qquad\qquad -x} & \text{Subtract } x \text{ from both sides.} \\ \quad\; y > 5 - x & \text{slope-intercept form} \end{array}$$

Graph the line $y = 5 - x$.

Since points that are on the line are NOT solutions of $y > 5 - x$, make the line dashed.

Then shade the part of the coordinate plane in which the rest of the solutions of $y > 5 - x$ lie.

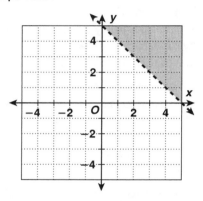

	Choose any point not on the line.
$(0, 0)$	
$y > 5 - x$	
$0 > 5 - 0$	Substitute 0 for x and 0 for y.
$0 > 5$	

Since $0 > 5$ is false, $(0, 0)$ is not a solution of $y > 5 - x$. Shade the side of the line that does not include the point $(0, 0)$.

Concepts covered in this chapter are very important for subsequent mathematic courses your child will be taking. Review these processes daily with your child.

Sincerely,

Holt Pre-Algebra

CHAPTER	**Family Letter**
11	*Linear Relationships*

Find each equation of direct variation, given that *y* varies directly with *x*.

1. *y* is 6 when *x* is 3 _____

2. *y* is 5 when *x* is 1 _____

3. *y* is 20 when *x* is 15 _____

4. *y* is 1 when *x* is −1 _____

5. *y* is −56 when *x* is 16 _____

6. *y* is 0 when *x* is 3 _____

Determine two solutions for the graph of each inequality.

7. _____

8. _____

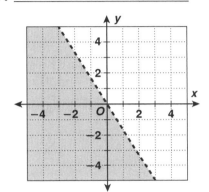

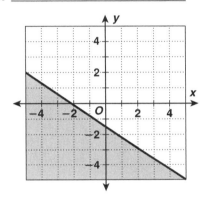

Determine if the boundary line is a solid line or dashed line.

9. $y > 3x - 4$ _____

10. $y \le 5x + 1$ _____

11. $x - y \ge 5$ _____

12. $x \le y + 7$ _____

13. $x < 7$ _____

14. $y \ge 6x - 1$ _____

Tell whether a line of best fit for each scatter plot would have a *positive* or *negative* slope. If a line of best fit would not be appropriate for the data, write *neither*.

15. _____

16. _____

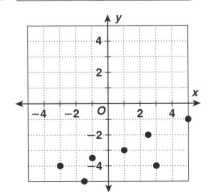

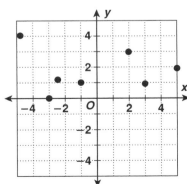

15. positive 16. neither

Answers: 1. $y = 2x$ 2. $y = 5x$ 3. $y = \frac{4}{3}x$ 4. $y = -x$ 5. $y = -\frac{7}{2}x$ 6. $y = 0$ 7. Possible answer: (−2, 2), (−1, −6) 8. Possible Answer: (0, −4), (−4, 1) 9. dashed 10. solid 11. solid 12. solid 13. dashed 14. solid

Holt Pre-Algebra

Name _____ Date _____ Class _____

Shading to Find the Shape

Directions
- Graph each of the linear inequalities on the coordinate system below.
- Use a dark pen or marker to outline the area that all linear inequalities have in common.
- Determine the shape created by the solutions of the linear inequalities.

Graph:

$y \leq 2$

$y \geq -2$

$y \geq x - 4$

$y \leq 4 - x$

$y \leq x + 3$

$y \geq -3 - x$

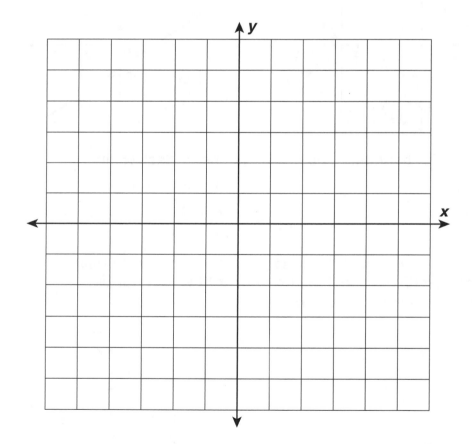

Answer: Hexagon

Holt Pre-Algebra

Family Letter

What We Are Learning

Sequences

Vocabulary
These are the math words we are learning:

Arithmetic Sequence An ordered list of numbers where the difference between consecutive terms is always the same.

Common Difference The difference between any two successive terms in an arithmetic sequence.

Common Ratio The ratio used to multiply each term to produce the next term in a geometric sequence.

Fibonacci Sequence An infinite sequence of numbers (1, 1, 2, 3, 5, 8, 13, ...). Starting with the third term, each number is the sum of the two previous numbers.

First Difference A sequence formed by subtracting each term of a sequence from the next term.

Geometric Sequence An ordered list of numbers that has a common ratio between consecutive terms.

Second Difference A sequence formed from differences of differences between terms of a sequence.

Sequence An ordered list of numbers or objects.

Dear Family,

Your child will be learning about **sequences** and how to find the pattern that defines the sequence. Your child will learn the difference between **geometric** and **arithmetic sequences,** and how to find **first** and **second differences** between terms.

In this lesson your child will learn how to recognize arithmetic sequences and how to determine patterns in the data. A sequence is arithmetic if the difference between one **term** and the next is always the same. Here is an example of how your child will identify arithmetic sequences.

Determine if each sequence could be arithmetic. If so, give the common difference.

A. 12, 15, 18, 21, 24, ...

12, 15, 18, 21, 24, ...   Subtract each term from the term
before it.
  3   3   3   3

The sequence could be arithmetic with a common difference of 3.

B. 13, 14, 16, 19, 23, ...

13, 14, 16, 19, 23, ...   Subtract each term from the term
before it.
  1   2   3   4

The sequence is not arithmetic.

In this lesson your child will learn how to recognize geometric sequences and how to determine patterns in data. A sequence is geometric if there is a common ratio between each term.

Determine if each sequence could be geometric. If so, give the common ratio.

3, −6, 12, −24, 48, ...

3, −6, 12, −24, 48, ...   Divide each term by the term
before it.
−2 −2 −2 −2

The sequence could be geometric with a common ratio of −2.

Holt Pre-Algebra

Term An element or number in a sequence.

For sequences that are not arithmetic or geometric, first and second differences can be used to find the pattern and other terms in the sequence.

If the first difference is not the same for all terms, then find the second difference. If this is a common difference, use it to work backward to find the terms in the sequence.

Use first and second differences to find the next three terms in the sequence.

5, 10, 20, 35, 55…

Sequence	5	10	20	35	55	80	110	145
1st Differences	5	10	15	20	25	30	35	
2nd Differences		5	5	5	5	5	5	

The second differences is 5. Fill in the third row with 5's.

Then add 5 to 20 in the second row to get 25 and then add 5 to 25 to get 30, etc.

Next add 25 to 55 in the first row to get 80 and then add 30 to 80 to get 110, etc.

Sometimes the rule for a sequence is given instead of the first couple of terms. To find the terms using the rule, substitute 1 for n and simplify. Then substitute 2 for n and simplify. Continue this process until you find all of the necessary terms.

Find the first five terms of the sequence defined by $a_n = n^2 + n + 1$.

$a_1 = 1^2 + 1 + 1 = 3$

$a_2 = 2^2 + 2 + 1 = 7$

$a_3 = 3^2 + 3 + 1 = 13$

$a_4 = 4^2 + 4 + 1 = 21$

$a_5 = 5^2 + 5 + 1 = 31$

The first five terms are 3, 7, 13, 21, 31.

Sequences have an important role in mathematics. Encourage your child to be aware of geometric and arithmetic sequences in everyday situations.

Sincerely,

Holt Pre-Algebra

Name _____ Date _____ Class _____

Family Letter
Sequences

Determine if each sequence could be arithmetic. If so, give the common difference.

1. −3, −1, 1, 3, 5, 7, ...

2. 3, $\frac{1}{3}$, $\frac{1}{9}$, $\frac{1}{27}$, ...

3. −25, −24, −22, −19, ...

_____   _____   _____

Find the given term in each arithmetic sequence.

4. 12th term: 1, 4, 7, 10, 13, ...

5. 36th term: 3, 7, 11, 15, 19, ...

_____   _____

Determine if each sequence could be geometric. If so, give the common ratio.

6. 12, 6, 3, 1.5, ...

7. 6, −12, 24, −48, ...

8. 3, 5, 7, 9, ...

_____   _____   _____

Find the given term in each geometric sequence.

9. 14th term: 3, 9, 27, 81, ...

10. 200th term: 5, −5, 5, −5, ...

_____   _____

Use first and second differences to find the next three terms in each sequence.

11. 13, 13, 16, 25, 43, 73, 118, ...

12. 4, 9, 29, 64, 114, 179, ...

_____   _____

Find the first five terms of each sequence defined by the given rule.

13. $a_n = \dfrac{n(n + 1)}{(n + 2)}$

14. $a_n = n^2(n - 1) + 2n$

_____   _____

Answers: 1. yes; 2 **2.** no **3.** no **4.** 34 **5.** 143 **6.** yes; $\frac{1}{2}$ **7.** yes, −2 **8.** no **9.** 4,782,969 **10.** −5 **11.** 181, 265, 373 **12.** 259, 354, 464 **13.** $\frac{2}{3}$, 1.5, 2.4, 3.3, 4.29 **14.** 2, 8, 24, 56, 110

Holt Pre-Algebra

Name _____ Date _____ Class _____

Family Fun
Seek and Find Sequences

Find the words in the grid. When you are done, the unused letters in the top part of the grid will spell out a hidden message. Pick them out from left to right, top line to bottom line. Words can go horizontally, vertically and diagonally in all eight directions.

```
E  C  N  E  R  E  F  F  I  D  D  N  O  C  E  S  M  E
A  R  I  T  H  M  E  T  I  C  S  E  Q  U  E  N  C  E
A  T  H  W  O  U  L  D  N  O  T  B  E  S  O  N  D  C
I  F  F  I  C  U  L  T  I  F  I  T  D  I  E  D  F  O
N  O  T  H  A  V  E  S  O  M  A  N  Y  U  N  U  I  M
M  B  E  R  S  L  D  K  Q  W  R  S  Q  J  Z  D  R  M
C  B  B  M  X  M  J  T  G  G  F  E  K  M  K  L  S  O
K  O  T  M  B  G  N  M  C  K  S  N  M  H  K  C  T  N
R  Y  M  Q  F  M  R  M  K  C  M  B  F  E  L  V  D  D
L  W  Z  M  Q  Z  R  X  I  F  Z  N  Y  C  L  V  I  I
W  N  R  H  O  E  M  R  T  F  R  V  L  N  M  W  F  F
M  D  K  B  T  N  T  K  M  K  L  B  W  E  Q  T  F  F
P  M  G  L  T  E  R  K  M  K  H  B  N  U  L  F  E  E
L  J  L  Y  M  Y  P  A  G  M  T  P  Q  Q  M  V  R  R
X  K  N  O  M  F  Q  M  T  N  Y  Z  H  E  T  P  E  E
W  G  E  K  Z  L  Z  T  G  I  K  R  Z  S  M  H  N  N
K  G  G  F  X  R  W  Z  J  L  O  T  N  M  V  X  C  C
F  I  B  O  N  A  C  C  I  S  E  Q  U  E  N  C  E  E
```

ARITHMETIC SEQUENCE   FIRST DIFFERENCE

COMMON DIFFERENCE   GEOMETRIC SEQUENCE

COMMON RATIO   SECOND DIFFERENCE

FIBONACCI SEQUENCE   SEQUENCE

Hidden message: Math would not be so difficult if it did not have so many numbers

Holt Pre-Algebra

Family Letter

Section B

What We Are Learning

Functions

Vocabulary
These are the math words we are learning:

Domain The set of all possible input values of a function.

Exponential Decay In an exponential function, when the output $f(x)$ gets smaller as the input (x) gets larger.

Exponential Function A function of the form $f(x) = p \cdot a^x$.

Exponential Growth In an exponential function, when the output $f(x)$ gets larger as the input (x) gets larger.

Function A rule that relates two quantities so that each input value corresponds to exactly one output value.

Function Notation The notation $f(x)$ used to describe a function.

Input The value substituted into a function.

Inverse Variation A relationship in which one variable quantity increases as another variable quantity decreases.

Linear Function A function whose graph is a non vertical–line.

Output The value that results when an input is substituted into a function.

Dear Family,

In this section, your child will learn about various types of **functions.** Your child will learn how to differentiate between the different types of functions, and how to graph each function type.

The two most popular ways to represent a function is with either a table or a graph, as shown in this example.

Make a table and graph of $y = x^2 + 3$.
Make a table of inputs and outputs. Use the table to make a graph.

x	$x^2 + 3$	y
-3	$(-3)^2 + 3$	12
-2	$(-2)^2 + 3$	7
-1	$(-1)^2 + 3$	4
0	$(0)^2 + 3$	3
1	$(1)^2 + 3$	4
2	$(2)^2 + 3$	7
3	$(3)^2 + 3$	12

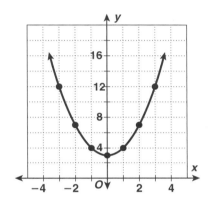

In this section your child will learn the difference between **exponential growth** and **decay.** When a function grows or decays exponentially, it means that the function is either growing or declining at a quick rate.

A certain element has a half-life of 14 hours, which means it takes 14 hours for half of the substance to decompose. Find the amount remaining from a 520 g sample after 112 hours.

$f(x) = p \cdot a^x$ x is the number of half-lives, and $f(x)$ is the amount of the substance left after x half-lives.

$f(x) = 520 \cdot a^x$ p is the original amount of the substance.

$f(x) = 520 \cdot \left(\frac{1}{2}\right)^x$ a is $\frac{1}{2}$ because the half-life is given.

To find the number of half-lives, x, divide 112 hours by

14 hours: $\frac{112}{14} = 8$ half-lives.

Now, find $f(8)$.

$f(8) = 520 \cdot \left(\frac{1}{2}\right)^8 = 2.03125$

There is approximately 2.03125 g left after 112 hours.

Holt Pre-Algebra

Parabola The basic shape of all quadratic functions.

Quadratic Function A function of the form $y = ax^2 + bx + c$, where $a \neq 0$.

Range The set of all possible output values of a function.

Your child will learn that the graph of all quadratic functions have the same shape, a parabola, and are typically represented by the function: $f(x) = ax^2 + bx + c$.

Create a table for the quadratic function and use it to make a graph.

$f(x) = x^2 - 1$

x	$f(x) = x^2 - 1$	y
-3	$(-3)^2 - 1$	8
-2	$(-2)^2 - 1$	3
-1	$(-1)^2 - 1$	0
0	$(0)^2 - 1$	-1
1	$(1)^2 - 1$	0
2	$(2)^2 - 1$	3
3	$(3)^2 - 1$	8

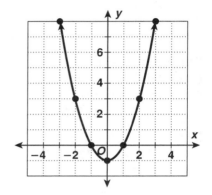

Plot the points and connect them with a smooth curve. This curve is called a parabola.

Students will also be learning how to tell if a relationship is an inverse variation, as shown below.

The table shows the amount of water needed to put out a fire over an area. Determine if the relationship is an inverse variation.

Gallons of water poured	3	6	12	24	48
Burned acres	150	75	37.5	18.75	9.375

$3(150) = 450$

$6(75) = 450$

$12(37.5) = 450$

$24(18.75) = 450$

$48(9.375) = 450$

The product of x and y is always the same, 450.

The relationship is an inverse variation: $y = \dfrac{450}{x}$.

The concepts introduced in this chapter are important for the next level of mathematics that your child will encounter. Help to keep these ideas fresh in your child's mind by helping them with their homework.

Sincerely,

Holt Pre-Algebra

CHAPTER 12 **Family Letter**
Functions

For each function, find $f(1)$, $f(-2)$, and $f(4)$.

1. _____

x	y
−2	6
0	8
1	10
4	18

2. _____

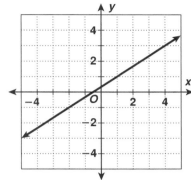

3. _____

$y = -1.2x + 4.1$

Write the rule for each linear function.

4. _____

x	−1	0	1	2
y	−5	−3	−1	1

5. _____

x	−2	0	2	4
y	4	3	2	1

6. _____

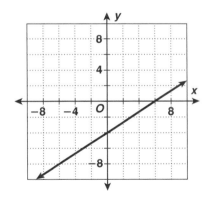

7. _____

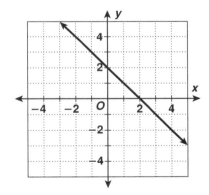

For each exponential function, find $f(-3)$, $f(0)$, and $f(3)$.

8. $f(x) = 0.5^x$

9. $f(x) = 3^x$

10. $f(x) = (-2)^x$

_____ _____ _____

Answers: 1. 10, 6, 18 2. 1, −1, 3 3. 2.9, 6.5, −0.7 4. $y = 2x − 3$ 5. $y = -\frac{1}{2}x + 3$ 6. $y = \frac{2}{3}x − 4$
7. $y = −x + 2$ 8. 8, 1, 0.125 9. 0.037, 1, 27 10. −0.125, 1, −8

Holt Pre-Algebra

CHAPTER
12

Family Fun
Function Tic-Tac-Toe

Directions

• Determine if each function is linear, exponential, or quadratic.

• Whoever answers the problem first can take the space and place their
mark X or O.

• The person with the most X's or O's wins.

<table>
<tr>
<td>

x	y
−2	4
−1	1
0	0
1	1
2	4
3	9

</td>
<td>

x	y
−2	3
−1	4
0	5
1	6
2	7
3	8

</td>
<td>

x	y
−2	$\frac{1}{4}$
−1	$\frac{1}{2}$
0	1
1	2
2	4
3	8

</td>
</tr>
<tr>
<td>

x	y
−4	−3
−2	1
0	5
2	9
4	13
6	17

</td>
<td>

free

</td>
<td>

x	y
−4	16
−2	6
0	4
2	10
4	24

</td>
</tr>
<tr>
<td>

x	y
−6	33
−4	13
−2	1
0	−3
2	1
4	13

</td>
<td>

x	y
0	2
1	6
2	18
3	54
4	162
5	486

</td>
<td>

x	y
−4	81
−2	9
0	1
2	9
4	81
6	729

</td>
</tr>
</table>

Answers: From left to right across top: quadratic, linear, exponential, middle: linear, exponential, quadratic, bottom: quadratic, exponential, exponential

Holt Pre-Algebra

Family Letter

What We Are Learning

Identifying and Simplifying Polynomials

Vocabulary
These are the math words we are learning:

Binomial A polynomial with 2 terms.

Degree of a Polynomial The degree of the term with the greatest degree.

Monomial A number or a product of numbers and variables with exponents that are whole numbers.

Polynomial One monomial or the sum or difference of polynomials.

Trinomial A polynomial with 3 terms.

Dear Family,

In this section, your child will learn about **polynomials:** how to identify them and how to simplify them. Your child will learn the differences between **monomials, binomials,** and **trinomials,** and how to classify polynomials by **degree.**

Your child will learn how to determine whether an expression is a monomial. The following are two examples of the process.

Determine whether the expression is a monomial.

A. $5x^4y^3$ *Yes, it is a monomial. 5 is a number; x^4 and y^3 are variables whose exponents are whole numbers.*

B. $\frac{1}{2}ab^{1.3}$ *No, it is not a monomial. $\frac{1}{2}$ is a number, but the exponent 1.3 is not a whole number.*

Here are some examples that further show the distinction between terms that are and are not monomials.

Monomials	$5t$, x^5, $2a^2b^5$, 8
Not monomials	$y^{3.4}$, 3^x, $\sqrt{y}$, $\frac{8}{c^2}$

In this section your child will also learn how to classify polynomials by the number of terms they contain.

Classify each expression as a monomial, a binomial, a trinomial, or not a polynomial.

A. $5.34m - 10.3n$ *It is a binomial; a polynomial with 2 terms.*

B. $-4c^2b$ *It is a monomial; a polynomial with 1 term.*

C. $12xy - 4x + 3y$ *It is a trinomial; a polynomial with 3 terms.*

D. $2x^2 + 4xy - \frac{4}{y}$ *Not a polynomial; there is a variable in the denominator.*

E. $\frac{1}{2}g^{2.8} + 7h$ *Not a polynomial; the exponent is not a whole number.*

Holt Pre-Algebra

Your child will also learn how to simplify polynomials by combining like terms. Two terms are like terms if they have the same variables raised to the same power.

In $5a^4b^6 + 2a^2b^6 + 4a^4b^6$, $5a^4b^6$ and $4a^4b^6$ are like terms because they have the same variables with the same powers. However, $5a^4b^6$ and $2a^2b^6$ are not like terms because a^4 is not raised to the same power as a^2.

Simplify polynomials by combining like terms.

$2x^4y^3 + 5x^2y^2 - 2 + 3x^4y^3$

$2x^4y^3 + 3x^4y^3 + 5x^2y^2 - 2$ *Arrange in descending order.*

$\widehat{2x^4y^3} + \widehat{3x^4y^3} + 5x^2y^2 - 2$ *Identify like terms.*

$5x^4y^3 + 5x^2y^2 - 2$ *Combine coefficients: $2 + 3 = 5$.*

Finally, your child will learn how to simplify polynomials by applying the Distributive Property. This means multiplying each of the terms in a polynomial that is inside parentheses by one or more terms outside the parentheses.

Simplify the polynomial by using the Distributive Property.

A. $2(3x^2 + 5x)$

$2(3x^2 + 5x)$ *Distributive Property*

$2 \cdot 3x^2 + 2 \cdot 5x$
$6x^2 + 10x$

B. $3(4g^2h - 2g) + 2g^2h + 1$

$3(4g^2h - 2g) + 2g^2h + 1$ *Distributive Property*

$3 \cdot 4g^2h - 3 \cdot 2g + 2g^2h + 1$

$12g^2h - 6g + 2g^2h + 1$ *Multiply.*

$12g^2h + 2g^2h - 6g + 1$ *Arrange in descending order.*

$14g^2h - 6g + 1$ *Combine like terms.*

Sincerely,

Holt Pre-Algebra

| CHAPTER | **Family Letter** |
| 13 | *Identifying and Simplifying Polynomials* |

Determine whether each expression is a monomial.

1. $10x^2y^4$ **2.** x^p **3.** $-\dfrac{1}{3}a^5b$ **4.** $3gt^{\frac{1}{2}}$

_____   _____   _____   _____

Classify each expression as a monomial, a binomial, a trinomial, or not a polynomial.

5. $-3m^2n^5$ **6.** $\dfrac{1}{5}x^{5.2} - 12y$ **7.** $8cd - 2c - 9d$ **8.** $2x^2 + 4xy$

_____   _____   _____   _____

Find the degree of each polynomial.

9. $x^3 + 4xy - 7$ **10.** $p + 4p^2 + 2$ **11.** $3 + 3a^3 - 2a^2$ **12.** $y - 4y^3 - 2y^5$

_____   _____   _____   _____

Identify the like terms in each polynomial.

13. $4a^2b - 2b + 3a^2 - 2a^2b$ **14.** $2m^2 + mn - m$ **15.** $6x^4y^3 - 3x^3y^4 + 2x^4y^3$

_____   _____   _____

Simplify the polynomial by combining like terms.

16. $3x^3 - 4x^2 + 2 - x^3 + 5x^2$ **17.** $7a^2b^3 + 4a^2b - 2b^2a - 2a^2b - 5a^2b^3$

_____   _____

Simplify the polynomial by using the distributive property.

18. $4(3m^2 - 2)$ **19.** $2(2x^2 + 3x) - x^2 - 8x$

_____   _____

Answers: 1. yes. **2.** no. **3.** yes. **4.** no. **5.** monomial. **6.** not a polynomial. **7.** trinomial. **8.** binomial. **9.** 3. **10.** 2. **11.** 3. **12.** 5. **13.** $4a^2b$ and $2a^2b$. **14.** no like terms. **15.** $6x^4y^3$ and $2x^4y^3$. **16.** $2x^3 + x^2 + 2$. **17.** $2a^2b^3 + 2a^2b - 2b^2a$. **18.** $12m^2$. **19.** $3x^2 - 2x$.

Holt Pre-Algebra

CHAPTER 13

Family Fun
Polynomial Derby

The collection of terms below includes three polynomials and their simplified form. See who can find them the quickest.

$3x^2 - 3xy + 1$	$10x^2 - 30$	$x^2 - 2x - 1$
$2(3x^2 - 10) + x^2 - 5$	$x^2 - 3xy$	$7x^2 - 2x - 1$
$-7x^2 - 2x - 1$	$-4xy + 3x^2 - 1 - 2x^2 + xy$	$5x^2 - 4$
$4x^2 - 5$	$3(2x^2 - 10) + x^2 + 5$	$x^2 - 4xy - 1$
$8x^2 + 2x - 1 - x^2$	$7x^2 - 2x - 1$	$7x^2 - 2x - 2$
$6x^2 - 20$	$7x^2 + 2x - 1$	$7x^2 + 15$
$2x^2 - 3xy - 1$	$8x^2 + 2x - 2 - 2x^2$	$2(3x^2 - 10) + x^2 + 5$
$x^2 - 3xy - 1$	$7x^2 - 15$	$2(3x^2 - 5) + x^2 + 5$

Holt Pre-Algebra

Family Letter

What We Are Learning

Adding, Subtracting, and Multiplying Polynomials

Vocabulary
These are the math words we are learning:

FOIL An easy way to remember to do the four multiplications needed when multiplying two binomials: the "First" terms, the "Outer" terms, the "Inner" terms, and the "Last" terms of the binomials.

Dear Family,

In this section, your child will learn about adding and subtracting polynomials. Your child will also learn how to multiply and divide polynomials by binomials. In learning how to multiply two binomials, your child will learn the **FOIL** method.

The Associative Property states
$a + b + c = (a + b) + c = a + (b + c)$. You apply this property when adding polynomials horizontally.

Add the polynomials horizontally.

$(4a^2b - 3ab - 2) + (8ab + 6a^2b + 5)$

$4a^2b - 3ab - 2 + 8ab + 6a^2b + 5$	*Associative Property*
$4a^2b + 6a^2b - 3ab + 8ab - 2 + 5$	*Arrange in descending order.*
$10a^2b + 5ab + 3$	*Combine like terms.*

Polynomials can also be added vertically. The important thing here is to keep like terms in the same column.

Add the polynomials vertically.

$(4x^2y - 3x + 4y) + (2x^2y + 6x + 3)$

$$4x^2y - 3x + 4y \qquad \textit{Place like terms in columns.}$$

$$+ \quad 2x^2y + 6x \qquad + 3$$
$$\overline{ 6x^2y + 3x + 4y + 3} \qquad \textit{Combine like terms.}$$

In this section, your child will also learn how to subtract polynomials. Subtraction is the opposite of addition. Therefore, your child first needs to be able to find the opposite of a polynomial.

Find the opposite of each polynomial.

$-6x^2 - 2$

$-(-6x^2 - 2)$

$6x^2 + 2 \qquad \textit{Remove the parentheses and distribute the sign.}$

Holt Pre-Algebra

Subtract the polynomials horizontally.

$(a^4 + 2a^3 - 2a) - (6a - 2a^4 + 3)$

$(a^4 + 2a^3 - 2a) + (-6a + 2a^4 - 3)$ *Add the opposite.*

$a^4 + 2a^3 - 2a - 6a + 2a^4 - 3$ *Apply the Associative Property.*

$3a^4 + 2a^3 - 8a - 3$ *Combine like terms.*

You child will learn how to multiply two monomials and a polynomial by a monomial. In both cases, when you multiply two powers of the same base, you add the exponents. To multiply two monomials, multiply the coefficients and add the exponents. To multiply a polynomial by a monomial, use the Distributive Property.

Multiply the polynomial by the monomial.

$5x^3y^2(2x^2y^4 - 3x^5y^2)$

$5x^3y^2(2x^2y^4 - 3x^5y^2)$ *Multiply each term in the parentheses by $5x^3y^2$.*

$10x^5y^6 - 15x^8y^4$ *Be sure to add exponents of the same base.*

To multiply two binomials, use the **FOIL** method. Multiply the "First" terms, then the "Outer" terms, then the "Inner" terms, then the "Last" terms.

Multiply the binomials.

$(x + 3)(x - 2)$

$(x + 3)(x - 2)$ *FOIL*

$x^2 - 2x + 3x - 6$ *Arrange in descending order.*

$x^2 + x - 6$ *Combine like terms.*

Sincerely,

Holt Pre-Algebra

CHAPTER **Family Letter**
13 *Adding, Subtracting, and Multiplying Polynomials*

Add the polynomials.

1. $(8a^2 + 4a - 2) + (a^2 - 2a + 1)$

2. $(-4x^2y + 3xy - 2) + (-5x^2y - 4xy + 2)$

_____

_____

Find the opposite of each polynomial.

3. $2ab^2c^3$

4. $-4x^3y$

5. $-8g^2h + 4gh - 3$

_____

_____

_____

Subtract the polynomials.

6. $(4a^2b + 3ab - 2) - (6a^2b + 4ab + 4)$

7. $(2x^2y^2 - 3xy + 5x) - (x^2y^2 + 2x - 3)$

_____

_____

Multiply.

8. $3a^2b(4a^3b^4 - 2ab^5)$

9. $2x^2y\,(x^3y^2 - 3yt + 2xp)$

_____

_____

Multiply.

10. $(4m + 1)(m - 5)$

11. $(2g - h)(3s - r)$

12. $(3x - 2y)(x - y)$

_____

_____

_____

13. $(x - 5)^2$

14. $(y + 3)(y - 3)$

15. $(g + h)^2$

_____

_____

_____

Answers: 1. $9a^2 + 2a - 1$. **2.** $-9x^2y - xy$. **3.** $-2ab^2c^3$. **4.** $4x^3y$. **5.** $8g^2h - 4gh + 3$.
6. $-2a^2b - ab - 6$. **7.** $x^2y^2 - 3xy + 3x + 3$. **8.** $12a^5b^5 - 6a^3b^6$. **9.** $2x^5y^3 - 6x^2y^2t + 4x^3yp$.
10. $4m^2 - 19m - 5$. **11.** $6gs - 2gr - 3sh + hr$. **12.** $3x^2 - 5xy + 2y^2$. **13.** $x^2 - 10x + 25$. **14.** $y^2 - 9$.
15. $g^2 + 2gh + h^2$.

Holt Pre-Algebra

CHAPTER 13 **Family Fun**
Head to Head

The problems and solutions below have missing elements. Face off against an opponent and see who can find all the missing elements first. Be sure to initial your answer to claim it. One point for each problem completed correctly and minus two points for each problem completed incorrectly.

$(4x^2 - 2x - 3) + ($ _____ $+ 12) = -3x^2 + x - 9$

$($ ___ $- 2x$ ___ $+ 3) - (-x^2$ ___ y ___ $) = -3x^2 - 6xy + 5$

$(3xy - 3x + 4$ ___ $) - (3x^2$ ___ $- y) = -3x^2 + 3xy - x + 5y$

2 ___ $y^3z^4($ ___ x^5 ___ $z^2) = 8x^7y^7z^6$

$(4x - 2)($ _____ $) = 28x^2 - 26x + 6$

$(-5x$ _____ $)(-2x$ ___ $) = 10x^2 - 11xy + 3y^2$

$($ _____ $)(3x - 2y) = 9x^2 - 4y^2$

$($ _____ $)^2 = 4x^2 - 6xy + 9y^2$

Answers:

$(4x^2 - 2x - 3) + (-7x^2 + 3x + 12) = -3x^2 + x - 36$

$(4x^2 - 2xy + 3) - (-x^2 + 4xy - 2) = -3x^2 - 6xy + 5$

$(3xy - 3x + 4y) - (3x^2 - 2x - y) = -3x^2 + 3xy - x + 5y$

$2x^2y^3z^4(4x^5y^4z^2) = 8x^7y^7z^6$

$(4x - 2)(7x - 3) = 28x^2 - 26x + 6$

$(-5x + 3y)(-2x + y) = 10x^2 - 11xy + 3y^2$

$(3x + 2y)(3x - 2y) = 9x^2 - 4y^2$

$(2x - 3y)^2 = 4x^2 - 6xy + 9y^2$

Holt Pre-Algebra

What We Are Learning

Sets

Vocabulary
These are the math words we are learning:

Element An object in a set.

Empty Set The set with no elements, also called the *null* set.

Finite Set A set containing a finite, or countable, number of elements.

Infinite Set A set containing an infinite, or uncountable, number of elements.

Intersection The intersection of sets *A* and *B* is the set of all elements that are in both *A* and *B*.

Set A collection of objects, called elements.

Subset Set *A* is a subset of set *B* if every element in *A* is also in *B*.

Union The union of sets A and B is the set of all elements that are in either A or B.

Dear Family,

Your child will be learning about **sets** and how to determine what **elements** belong to a set. Your child will be able to identify types of sets such as **finite set, infinite set,** and **empty set.** Your child will also learn how to identify the elements in a **union** or **intersection** of sets, as well as elements completely contained as a **subset** of another set.

In the first part of this section, your child will learn how to identify elements of a set, subsets, and finite and infinite sets. Your child will also learn how to use special symbols. The first three are: $\in$, which is read as "is an element of"; $\notin$, which is read as "is not an element of"; and { } which is read as "the set of all."

Insert the correct symbol to make each statement true.

A. 2 _____ {odd numbers}   2 is not an odd number.

   2 $\notin$ {odd numbers}

B. celery _____ {green vegetables}   Celery is a green vegetable.

   celery $\in$ {green vegetables}

Your child will learn to use the symbol $\subset$ for "is a subset of" and $\not\subset$ for "is not a subset of."

Determine whether the first set is a subset of the second set. Use the correct symbol.

A. Q = {fractions} R = {real numbers} Every fraction is a real number.

   Yes, $Q \subset R$

B. C = {cows} R = {fish}   A cow is a mammal and not a fish.

   No, $Q \not\subset R$

Your child will learn to identify finite and infinite sets.

Tell whether each set is finite or infinite.

A. {integers from 1 to 100}   There are exactly 100 integers in the set.

   finite

B. {integers over 100}   The number of integers over 100 is infinite.

   infinite

Holt Pre-Algebra

In this section your child will also learn about the intersection and union of sets and learn to use three symbols as follows: $A \cap B$ is read as "the intersection of set A and set B"; $A \cup B$ is read as "the union of set A and set B"; and { } or $\emptyset$ means the empty set or *null set*.

Find the intersection of the sets.

A. $P = \{1, 3, 8, 12\}$ $T = \{1, 8, 14, 20\}$
  $P \cap T = \{1, 8\}$

The only elements that appear in both P and T are 1 and 8.

B. $Q = \{$positive fractions$\}$
  $R = \{$negative integers$\}$
  $Q \cap R = \emptyset$ or { }

There are no positive fractions that are also negative integers.

Find the union of the sets.

A. $Q = \{1, 3, 8, 12\}$ $R = \{1, 8, 14, 20\}$
  $Q \cup R = \{1, 3, 8, 12, 14, 20\}$

1, 3, 8, 12, 14, and 20 are either in Q or R.

B. $L = \{x \mid x > 3\}$ $G = \{x \mid x < 20\}$
  $\{x \mid x > 3\}$ is read as "all x such that $x > 3$."
  $L \cup G = \{$real numbers$\}$

Every real number is either greater than 3 or less than 20.

In this section your child will also learn about Venn diagrams. A Venn diagram is made up of overlapping circles and shows relationships between sets.

Use the Venn diagram to identify intersections, unions, and subsets.

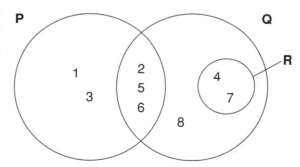

A. Intersection: $P \cap Q = \{2, 5, 6\}$

2, 5, and 6 are elements in P and Q.

B. Union: $P \cup Q = \{1, 2, 3, 4, 5, 6, 7, 8\}$

1, 2, 3, 4, 5, 6, 7 and 8 are elements in P or Q.

C. Subset: $R \subset Q$

All elements in R are elements in Q.

Sincerely,

Holt Pre-Algebra

Name _____ Date _____ Class _____

Family Letter
Sets

Insert the correct symbol (∈ or ∉) to make each statement true.

1. 16 ____ {multiples of 4}   **2.** △ ____ { trapezoids}   **3.** shark ____ {dinosaurs}

Determine whether the first set is a subset of the second set. Use the correct symbol (⊂ or ⊄).

4. *D* = {decimals} *R* = {real numbers}   **5.** *P* = {parallelograms} *R* = {rectangles}

_____   _____

Tell whether each set is finite or infinite.

6. {hairs on your head}   **7.** {integers divisible by 10}   **8.** {fractions equal to 4}

_____   _____   _____

Find the intersection of the sets.

9. *F* = {5, 6, 7, 8} *G* = {6, 8, 9, 10}   **10.** *N* = {negative integers} *R* = {− 3, −1, 2, 23}

_____   _____

Find the union of the sets.

11. *Y* = {0, 3, 4, 8} *G* = {0, 4, 8, 9}   **12.** *D* = {odd integers} *E* = {even integers}

_____   _____

Identify intersections and unions in the Venn diagram.

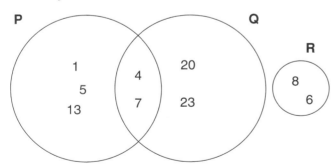

13. *P* ∪ *Q*   **14.** *P* ∩ *Q*   **15.** *P* ∩ *R*

_____   _____   _____

Family Fun
Set Rummy

The object of this game is to form true statements about sets shown in the Venn diagram below. To play this game, you need a collection of 24 cards (or slips of paper) numbered 1–6: 4 of them should be printed with a "1," 4 with a "2," and so on. The cards are arranged randomly in a pile.

A player draws 5 cards. After studying the cards and the Venn diagram, the player lays down a row of 4 or 5 cards in any order desired. The cards are then read as follows:

Position 1 (1 or 2 cards): the value of the card(s), 1–12
    Position 2 (1 card): 1 = A, 2 = B, 3 = C, 4 = D, 5 = E, 6 = F
   Position 3 (1 card): odd card = "∩"; even card = "∪"
   Position 4 (1 card): 1 = A, 2 = B, 3 = C, 4 = D, 5 = E, 6 = F

The cards are then read as: [Pos. 1] ∈ [Pos. 2] [Pos. 3] [Pos. 4]. So if the player were to lay down 5 − 2 − 1 − 3 − 2, he or she would then declare this as "7 is an element of the intersection of sets *A* and *B*." If everyone agrees that a statement is true, the player receives points: 2 points for an intersection statement and 1 point for a union statement. If a player cannot form a true statement, he or she passes, and the next player goes. The first player to get 10 points wins.

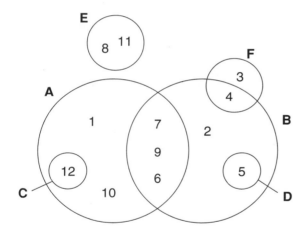

What We Are Learning

Deductive Reasoning and Networks

Vocabulary
These are the math words we are learning:

Circuit A path that ends at the same vertex at which it began and doesn't go through any edge more than once.

Compound Statement A statement formed by combining two or more simple statements.

Conclusion The statement *Q* in the conditional "If *P*, then *Q*."

Conditional See *if-then*.

Conjunction A compound statement of the form *P* and *Q*, where *P* and *Q* are simple statements.

Connected Graph A graph in which there is a path between every vertex and every other vertex.

Deductive Reasoning The application of a true conditional to a situation.

Degree (of a vertex) The number of edges touching a vertex.

Disjunction A compound statement of the form *P* or *Q*, where *P* and *Q* are simple statements.

Edge A line segment or arc joining two vertices.

Dear Family,

In this section your child will learn about **compound statements** and how to determine their **truth value** using a **truth table**.

In the first part of this section your child will use a truth table for a conjunction. The conjunction *P* and *Q* is true if both *P* <u>and</u> *Q* are true.

Make a truth table for the conjunction *P* and *Q*, where *P* is "Pat's car gets at least 30 miles per gallon" and *Q* is "Pat just spent $100 or more for repairs."

Example	*P*	*Q*	*P* and *Q*
Pat's car gets 32 miles per gallon and he just spent $150 for repairs.	True	True	True
Pat's car gets 35 miles per gallon and he just spent $85 for repairs.	True	False	False: only one of the conditions is met.
Pat's car gets 28 miles per gallon and he just spent $125 for repairs.	False	True	False: only one of the conditions is met.
Pat's car gets 29 miles per gallon and he just spent $95 for repairs.	False	False	False: neither condition is met.

In this section your child will also learn to identify the **hypothesis** and **conclusion** in a **conditional**.

Identify the hypothesis and conclusion in each conditional.

A. If I pass the exam, then I will graduate.
Hypothesis: I pass the exam.
Conclusion: I will graduate.

Identify the statements following if *and* then.

B. If a rhombus has four right angles, it is a square.
Hypothesis: A rhombus has four right angles.
Conclusion: It is a square.

The word then *may be omitted from conditional statements.*

Holt Pre-Algebra

Euler circuit A circuit that goes through every edge of a connected graph. (Pronounced *oiler*)

Graph In *graph theory,* a network of points and line segments or arcs that connect the points.

Hamiltonian Circuit A path that ends at the beginning vertex and passes through each of the other vertices in the graph exactly once.

Hypothesis The statement *P* in the conditional "If *P,* then *Q.*"

If-Then Statement A compound statement in the form "If *P,* then *Q.*" Also called a *conditional.*

Network The collection of points and line segments or arcs in a graph.

Path A way to get from one vertex to another along one or more edges.

Premise A conditional statement in a deductive argument.

Truth Value The truth value of a statement is either *true* or *false.*

Truth Table A way to show the truth value of a compound statement.

Vertex The point at which two edges meet.

In this section your child will also learn to use **deductive reasoning** to make a conclusion from a deductive argument.

Make a conclusion, if possible, from the deductive argument.

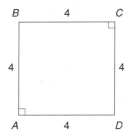

If ∠A and ∠C = 90° and the lengths of all four sides are equal, then quadrilateral *ABCD* is a square.
∠*A* and ∠*C* = 90°
The lengths of all four sides are equal.

Conclusion: Quadrilateral *ABCD* is a square.

The premises are true.
The conclusion is true.

Your child will also learn about **graphs** as understood in a branch of mathematics called *graph theory.* Your child will learn that the **degree** of a **vertex** equals the number of **edges** touching the vertex. He will also learn to determine whether a graph is a **connected graph.**

Find the degree of each vertex and determine whether the graph is connected.

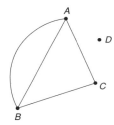

Vertex	Degree
A	3
B	3
C	2
D	0

The graph is not connected.

There is no path from vertex *D* to another vertex.

Finally, in this section your child will learn to find **Hamiltonian circuits** in a graph.

Find all the Hamiltonian circuits that begin at vertex *S.*

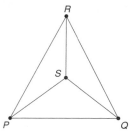

Each path begins and ends at *S*, and passes through all the other vertices exactly once.

SRPQS, SPQRS, SQRPS, SRQPS, SQPRS, and *SPRQS.*

Sincerely,

Holt Pre-Algebra

CHAPTER 14	**Family Letter**

Deductive Reasoning and Networks

1. Complete the truth table for the disjunction *P* or *Q*, where *P* is "It is warm" and *Q* is "It is windy."

   The disjunction *P* or *Q* is "It is warm or it is windy."

Example	P	Q	P or Q
It is warm. It is windy.	T	___	___
It is not warm. It is windy.	___	T	___
It is warm. It is not windy.	___	F	___
It is not warm. It is not windy.	F	___	___

Identify the hypothesis and the conclusion in each conditional.

2. If *x* = 5, then *x* is an odd integer.

   Hypothesis: _____

   Conclusion: _____

3. If I work hard, I will succeed.

   Hypothesis: _____

   Conclusion: _____

Make a conclusion, if possible from each deductive argument.

4. If *x* = 4, then 2*x* is an even integer.

   x = 13

   Conclusion: _____

5. If Ed is here, the party will be lively.

   Ed is here.

   Conclusion: _____

6. Find the degree of each vertex and determine whether the graph is connected.

   A _____ *B* _____ *C* _____

   D _____ *E* _____ *F* _____

   _____

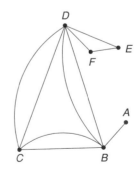

Answers: 1.

2. H: *x* = 5; C: *x* is an odd integer. 3. H: I work hard; C: I will succeed.
4. No conclusion can be made because the hypothesis is not true
(*x* is not equal to 4). 5. The party will be lively. 6. *A* = 1, *B* = 5, *C* = 4,
D = 6, *E* = 2, *F* = 2; connected.

P	Q	P or Q
True	True	True
False	True	True
True	False	True
False	False	False

Holt Pre-Algebra

| CHAPTER |
| 14 |

Family Fun
Head to Head

In studying about deductive reasoning and networks, you have learned a number of new terms with—like so many words in math—extremely precise meanings. See if you can rediscover these terms one more time using deductive reasoning as a tool. They appear in code below. Each letter of the regular alphabet has been replaced by a different letter. Azzh xtrp in your deductions!

1. H K H T R U G W K   B K D I Z E G E A
   _ _ _ _ _ _ _ _   _ _ _ _ _ _ _ _

2. U B T U S   U D L X K
   _ _ _ _ _   _ _ _ _ _

3. S N M Z U S K I G I
   _ _ _ _ _ _ _ _ _

4. R Z E R X T I G Z E
   _ _ _ _ _ _ _ _ _

5. R Z E E K R U K H   A B D M S
   _ _ _ _ _ _ _ _   _ _ _ _ _

6. E K U F Z B P
   _ _ _ _ _ _ _

7. K H A K
   _ _ _ _

8. U B T U S   W D X T K
   _ _ _ _ _   _ _ _ _ _

9. G V – U S K E   I U D U K Q K E U
   _ _   _ _ _ _   _ _ _ _ _ _ _ _ _

10. K T X K B   R G B R T G U
    _ _ _ _ _   _ _ _ _ _ _ _

11. W K B U K Y
    _ _ _ _ _ _

12. R Z E C T E R U G Z E
    _ _ _ _ _ _ _ _ _ _ _

13. R Z Q M Z T E H   I U D U K Q K E U
    _ _ _ _ _ _ _ _   _ _ _ _ _ _ _ _ _

14. R Z E H G U G Z E D X
    _ _ _ _ _ _ _ _ _ _ _

15. M B K Q G I K
    _ _ _ _ _ _ _

16. H G I C T E R U G Z E
    _ _ _ _ _ _ _ _ _ _ _

17. S D Q G X U Z E G D E   A B D M S
    _ _ _ _ _ _ _ _ _ _   _ _ _ _ _

18. M D U S
    _ _ _ _

Answers: 1. deductive reasoning. 2. truth table. 3. hypothesis. 4. conclusion. 5. connected graph. 6. network. 7. edge. 8. truth value. 9. if-then statement. 10. Euler circuit. 11. vertex. 12. conjunction. 13. compound statement. 14. conditional. 15. premise. 16. disjunction. 17. Hamiltonian graph. 18. path.

Holt Pre-Algebra